MATHEMATICAL EXPLORATIONS

African Institute of Mathematics Library Series

The African Institute of Mathematical Sciences (AIMS), founded in 2003 in Muizenberg, South Africa, provides a one-year postgraduate course in mathematical sciences for students throughout the continent of Africa. The **AIMS Library Series** is a series of short innovative texts, suitable for self-study, on the mathematical sciences and their applications in the broadest sense.

A complete list of books in the series can be found at www.cambridge.org/mathematics. Recent titles include the following:

Introduction to Atmospheric Modelling
DOUW G. STEYN

Mathematical Modelling in One Dimension
JACEK BANASIAK

A First Course in Computational Algebraic Geometry
WOLFRAM DECKER AND GERHARD PFISTER

Ordinary Differential Equations
BERND J. SCHROERS

From Measures to Itô Integrals
EKKEHARD KOPP

Creative Mathematics
ALAN F. BEARDON

AIMS Library Series

MATHEMATICAL EXPLORATIONS

ALAN F. BEARDON
University of Cambridge

CAMBRIDGE
UNIVERSITY PRESS

CAMBRIDGE
UNIVERSITY PRESS

University Printing House, Cambridge CB2 8BS, United Kingdom

Cambridge University Press is part of the University of Cambridge.

It furthers the University's mission by disseminating knowledge in the pursuit of education, learning, and research at the highest international levels of excellence.

www.cambridge.org
Information on this title: www.cambridge.org/9781316610565

First published 2016

Printed in the United Kingdom by Clays, St Ives plc

A catalogue record for this publication is available from the British Library.

ISBN 978-1-316-61056-5 Paperback

Contents

Preface	*page* ix
How to use this book	x

1 Paying for parking — 1
1.1 The problem — 1
1.2 Coprime values — 1
1.3 Which charges require change? — 2
1.4 Which charges cannot be paid? — 4
1.5 Three denominations of coins — 6

2 Lengths and angles — 7
2.1 The problem — 7
2.2 A generalisation — 9
2.3 Another generalisation — 10
2.4 Diophantine equations — 12
2.5 The value of π — 14

3 Magic squares — 17
3.1 Magic squares — 17
3.2 The space of 3×3 magic squares — 19
3.3 The space of 4×4 magic squares — 21
3.4 The dimension of the space of magic squares — 22
3.5 Lagrange's interpolation formula — 24

4 Intersecting chords — 26
4.1 Dividing a disc by chords — 26
4.2 The recurrence relation for R_n — 27
4.3 Lagrange's interpolation formula — 29

4.4	Euler's formula	31
4.5	Similar problems	32

5 Crossing squares   **34**
5.1	The problem	34
5.2	Lattice points on the diagonal D	35
5.3	The general solution	36
5.4	An alternative approach	37
5.5	A combinatorial proof	38
5.6	The problem in three dimensions	39

6 Repeated vector products   **42**
6.1	The problem	42
6.2	Vector products	42
6.3	A geometric approach	43
6.4	An algebraic approach	44
6.5	Quaternions	46

7 A rolling disc   **48**
7.1	A disc rolling in a tray	48
7.2	One circuit of the tray	49
7.3	Periodic motion	50
7.4	Non-periodic motion	51

8 Sums of powers of digits   **54**
8.1	The sum of the digits	54
8.2	The sum of the squares of the digits	56
8.3	A general result	58
8.4	Working in base B	58
8.5	The fixed points of f in base B	60

9 The metric dimension   **63**
9.1	A problem in robotics	63
9.2	The metric dimension of metric spaces	64
9.3	Bisectors	66
9.4	Cubes and hypercubes	68
9.5	Linear equations	70
9.6	A lower bound for $\beta(V^n)$	72

10 Primes and irreducible elements   **73**
| 10.1 | Primes and the integers | 73 |

10.2	Primes and units in a ring	73
10.3	Gaussian integers	75
10.4	Irreducible matrices	76
10.5	Irreducible polynomials	78
11	**The symmetries of a quadrilateral**	**81**
11.1	The problem	81
11.2	Quadrilaterals	82
11.3	Symmetries	82
11.4	Plane quadrilaterals	83
11.5	Quadrilaterals in $\mathbb{R}^3$	86
12	**Removing a vertex**	**90**
12.1	Removing the vertex of a cube	90
12.2	Two geometric solutions to Problem 12.1	91
12.3	An analytic solution	93
12.4	A generalisation	93
12.5	More general tetrahedra	95
13	**Squares within squares**	**96**
13.1	Squares on a geoboard	96
13.2	The solutions	96
13.3	Areas of squares on a geoboard	98
13.4	Areas of polygons on a geoboard	98
13.5	Pick's Theorem	100
13.6	Higher dimensions	102
13.7	Pick's Theorem in higher dimensions	103
14	**Catalan numbers**	**104**
14.1	Introduction	104
14.2	Binary operations	105
14.3	Paths joining lattice points	106
14.4	The number of positive paths on $[0, 2n]$	108
14.5	Photographs	110
14.6	An explicit formula for C_n	111
14.7	Why is $0! = 1$	113
	References	115
	Index	116

Preface

This book follows on from the book *Creative Mathematics* in this series which began with three essays (on research into, on writing about and on presenting mathematics) and then continued with a series of problems, each of which was divided into three parts. Part I provided an introduction to the problem followed by some elementary questions about the problem. Part II contained an answer to these questions, as well as a deeper discussion and a generalisation of the problem. This led to more advanced questions which were discussed in Part III.

This book is a natural development of this approach, the main purpose of which is to give the reader experience in working on (as far as the reader is concerned) unsolved problems. The problems in this book are, generally speaking, more difficult than those in *Creative Mathematics*, and we assume a greater level of maturity of the reader.

One of the main purposes of this book is to show that mathematical problems are often solved using mathematics that is not, at first sight, connected to the problem, and readers are encouraged (and even urged) to consider as wide a variety of mathematical ideas as possible when trying to solve a problem. The reader will no doubt have met problems in what might be called 'recreational mathematics' where problems are solved for amusement, without necessarily understanding or investigating the key mathematical ideas that lie behind the solution. Here, by contrast, we focus on the important underlying ideas rather than on the solution itself.

How to use this book

This book is written to help the reader learn how to do research in mathematics. Each chapter contains a project that has been chosen not because of its mathematical importance but because (in the view of the author) it provides a good illustration of how arguments develop, and how new questions arise once some progress is made. These projects have also been chosen because they do not require a deep mathematical background in order to understand the problem and start investigation. Nevertheless, the reader will probably have to learn some more mathematics in order to solve the problems. Some of the problems do not have easy answers, and some are not yet completely solved.

Each chapter focuses on one topic, and although some results and proofs are given in the discussion, many steps are omitted, and it is the responsibility of the reader to locate and fill these gaps. The general rule is that the reader should check every step and provide as much extra material as is necessary to ensure their complete understanding of each step. As we progress through a project, specific questions are asked, and the reader will need to interpret, or clarify, some of these before a solution is attempted. It hardly needs saying that the whole purpose of the book is that the reader should fully engage with these problems and fill in the (many) missing steps in the text itself. Although some theorems, and their proofs, are given, these do not have quite the same role as in most textbooks. The theorems given here serve the purpose of making further progress *in order that we can ask yet more questions*, for this is the real nature of research.

1

Paying for parking

1.1 The problem

Suppose that you have to pay for parking your car by putting coins in a machine that only accepts coins of values p and q units. *Which charges can you pay for without requiring any change?* The answer is obvious if $p = 1$, or $q = 1$, or $p = q$, so, from now on, *we shall assume that $p \geqslant 2, q \geqslant 2$ and $p \neq q$.* Clearly, the set of charges that you can pay for without requiring change is

$$M(p, q) = \{mp + nq : m, n = 0, 1, \ldots\},$$

and the problem is to say as much as you can about this set. From now on we shall omit the phrase 'without requiring change', although we shall always assume that this condition applies.

To begin, for a given pair (p, q), say $(3, 5)$, you might mark the points (mp, nq), $m, n = 0, 1, 2, \ldots$ on graph paper and see whether any ideas emerge from this picture.

1.2 Coprime values

It is obvious that if p and q are even then we can only pay an even number of units. More generally, if k is the greatest common divisor of p and q, that is, $\gcd(p, q) = k$, then we can only pay amounts that are integer multiples of k. These comments suggest that we proceed as follows. Let d be the greatest common divisor of p and q, and let $p_1 = p/d$ and $q_1 = q/d$. Then p_1 and q_1 are coprime integers, and (in the obvious sense)

1

$$M(p, q) = d \{mp_1 + nq_1 : m, n = 0, 1, \ldots\} = d\, M(p_1, q_1).$$

This argument shows that it is sufficient to consider $M(p_1, q_1)$ or, equivalently, to restrict ourselves to the case where p and q are coprime. Thus, from now on, *we shall assume that p and q are coprime*; that is, $\gcd(p, q) = 1$. How does this help? Well, it is extra information which we should be able to use to make further progress, *but only if we know some facts about coprime integers*. Thus we must now turn to number theory.

The most important consequence of $\gcd(p, q) = 1$ is that *there are integers r and s such that $rp + sq = 1$*. Briefly, we recall the proof. Consider the group $G = \{mp + nq : m, n \in \mathbb{Z}\}$, where $\mathbb{Z}$ is the set of integers. As $\mathbb{Z}$ is a cyclic group, and G is a subgroup of $\mathbb{Z}$, we see that G is cyclic. Thus we can write $G = \{kg : k \in \mathbb{Z}\}$, where $g > 0$, so that

$$\{mp + nq : m, n \in \mathbb{Z}\} = G = \{kg : k \in \mathbb{Z}\}.$$

It follows that there are integers k_1 and k_2 with $p = 1.p + 0.q = k_1 g$ and $q = 0.p + 1.q = k_2 g$. We deduce that g divides both p and q, so that $g = 1$, and then $\{mp + nq : m, n \in \mathbb{Z}\} = G = \mathbb{Z}$. As $1 \in \mathbb{Z}$, this shows that $mp + nq = 1$ for some m and n.

1.3 Which charges require change?

The reader should now carry out a few numerical experiments (on a computer), and these should suggest the following preliminary result.

Lemma 1.1 *Suppose that* $\gcd(p, q) = 1$. *Then there are only a finite number of charges which cannot be paid.*

Note that as the conclusion of Lemma 1.1 is false when p and q are not coprime, we will have to use the assumption that $\gcd(p, q) = 1$ somewhere in our proof. We shall illustrate the idea behind a general proof with a specific example, namely when $p = 5$ and $q = 11$. The first observation is that if we can pay five *consecutive* charges, say $N, N + 1, \ldots, N + 4$, then we can pay all amounts above N (simply by using more coins of value 5). Thus the problem is reduced to showing that we can pay five consecutive amounts.

Now as gcd(5, 11) = 1, we can solve $5m + 11n = 1$, say with $m = -2$ and $n = 1$ (there are other solutions here, for example, $m = 9$ and $n = -4$). This means that to pay an additional unit charge, we can put *one more* 11-*unit coin* in the machine and *two fewer* 5-*unit coins*. Of course, this is possible only if we have already put two (or more) 5-unit coins in the machine. Suppose that we can pay an amount N with, say a coins of 5 units and b coins of 11 units. Then $N = 5a + 11b$ and

$$N + 1 = 5a + 11b + \big(1.11 + (-2)5\big) = (a - 2)5 + (b + 1)11,$$
$$N + 2 = 5a + 11b + \big(2.11 + (-4)5\big) = (a - 4)5 + (b + 2)11,$$
$$N + 3 = 5a + 11b + \big(3.11 + (-6)5\big) = (a - 6)5 + (b + 3)11,$$
$$N + 4 = 5a + 11b + \big(4.11 + (-8)5\big) = (a - 8)5 + (b + 4)11.$$

Clearly, all these payments are possible if $a \geqslant 8$ and $b \geqslant 0$, so, for example, by taking $a = 8$ and $b = 0$ we see that it is possible to pay each of the amounts $40, 41, \ldots, 44$.

Problem 1.1 Generalise the argument given above to provide a proof of Lemma 1.1. The starting data is a pair of coprime, positive integers p and q, and the existence of integers u and v such that $pu + qv = 1$, where (necessarily) either $u < 0 < v$ or $v < 0 < u$. Now show that, providing N is sufficiently large, it is possible to pay each of the charges $N, N + 1, \ldots, N + p$. How large must N be (in terms of p and q) for this to be possible?

As is so often the case, a positive result raises further questions.

Problem 1.2 What is the largest amount that *cannot* be paid? How many different values *cannot* be paid?

Notice how our interest has changed from amounts that *can* be paid to amounts that *cannot* be paid. A change of emphasis is often crucial in solving problems.

Problem 1.3 The reader should now use a computer and show experimentally that the following result is true.

Theorem 1.2 *Suppose that* $\gcd(p, q) = 1$. *Then we cannot pay a charge of* $pq - p - q$, *but all charges above this amount can be paid.*

Proof First, we show that we cannot pay $pq - p - q$, and to do this we argue by contradiction. We suppose, then, that we can pay this amount; thus we assume that there are non-negative integers u and v such that $pq - p - q = up + vq$, or

$$pq = (u + 1)p + (v + 1)q.$$

Since $\gcd(p, q) = 1$, this shows that p divides $v + 1$, so that $v + 1 \geqslant p$. Likewise, $u + 1 \geqslant q$. Thus $pq = (u + 1)p + (v + 1)q \geqslant 2pq$, which is false. This contradiction shows that we cannot pay $pq - p - q$ units.

In order to show that all charges greater than $pq - p - q$ can be paid, we choose any integer n with $n > pq - p - q$. Next, as $\gcd(p, q) = 1$, there are integers a and b such that $pa + bq = 1$; thus $(na)p + (nb)q = n$. Now we can always write na as a multiple of q plus a remainder, say $na = cq + d$, where $0 \leqslant d < q$. Then

$$pq - p - q < n = (na)p + (nb)q$$
$$= pd + (pc + nb)q$$
$$\leqslant p(q - 1) + (pc + nb)q.$$

This shows that $pc + nb > -1$ so that $pc + nb \geqslant 0$. Since $n = pd + (pc + nb)q$, it follows that a charge of n can be paid (with d coins of value p and $pc + nb$ coins of value q). □

1.4 Which charges cannot be paid?

We now ask *how many* charges cannot be paid?

Problem 1.4 The reader should provide numerical examples to show *experimentally* that exactly $\frac{1}{2}(p - 1)(q - 1)$ charges cannot be paid. Note that as $\gcd(p, q) = 1$, one of p and q must be odd, so that $(p - 1)(q - 1)$ is always an even integer.

Theorem 1.3 *Suppose that* $\gcd(p, q) = 1$. *Then* $\frac{1}{2}(p - 1)(q - 1)$ *charges cannot be paid.*

Proof We begin by considering the number $N(k)$ of ways that we can pay for a charge k. Clearly, $N(k)$ is the number of pairs (u, v) of non-negative integers u and v with $pu + qv = k$. As there are integers a and

b with $pa + qb = 1$, there are certainly integers u ($= ka$) and v ($= kb$) such that $pu + qv = k$. It is well known that this means that the general solution in integers of $px + qy = k$ is given by $(x, y) = (u + qt, v - pt)$, where $t \in \mathbb{Z}$. These points lie on the line L_k given by $px + qy = k$ in $\mathbb{R}^2$ and are equally spaced along L_k, with two consecutive points being a distance $\sqrt{p^2 + q^2}$ apart (the reader should draw a diagram here). Now the line L_k meets the region $\{(x, y) : x \geqslant 0, y \geqslant 0\}$ in the segment, say S_k, with endpoints $(k/p, 0)$ and $(0, k/q)$, and S_k has length $(k/pq)\sqrt{p^2 + q^2}$. It follows that if $k < pq$ then there is at most one point (x, y) with integer co-ordinates on S_k; hence *at most one way that we can pay a charge k*. We leave the reader to show that there are exactly two ways to pay a charge of pq (namely q coins of value p, or p coins of value q), and these correspond to the points $(q, 0)$ and $(0, p)$ which are, in fact, the endpoints of the seqment S_{pq}.

A *lattice point* is a point in $\mathbb{R}^2$ that has integer co-ordinates, and the problem has now been reduced to counting lattice points. Let T be the triangular region given by $x \geqslant 0$, $y \geqslant 0$ and $px + qy < pq$, and let $|T|$ denote the number of lattice points in T. We have just seen that there are exactly $|T|$ values of k in $\{0, 1, 2, \ldots, pq - 1\}$ that can be paid; hence (since all values above $pq - 1$ can be paid), *there are exactly $pq - |T|$ non-negative values k which cannot be paid*. It remains to find $|T|$.

Now consider the triangular region T^* given by the inequalities $x \leqslant q$, $y \leqslant p$ and $px + qy > pq$ (the reader is advised to draw a diagram). Then the union of the three mutually disjoint sets T, S_{pq} and T^* is the rectangle R given by $0 \leqslant x \leqslant q$ and $0 \leqslant y \leqslant p$. If we now denote the number of lattice points in a set E by $|E|$, we have

$$|T| + 2 + |T^*| = |T| + |S_{pq}| + |T^*| = |R| = (p + 1)(q + 1).$$

The last step in the argument is to show that $|T| = |T^*|$; this can be done by considering the map $(x, y) \mapsto (q - x, p - y)$, and we leave this for the reader to complete. With this we have

$$2|T| + 2 = (p + 1)(q + 1),$$

which implies that $pq - |T| = \frac{1}{2}(p - 1)(q - 1)$ as required. $\quad\square$

Problem 1.5 Given p and q, is it possible to say anything about the actual amounts that cannot be paid?

1.5 Three denominations of coins

We now ask about the situation when we are allowed to pay using coins of *three* values p, q and r. Since we can pay any charge that exceeds $(p-1)(q-1)$ by using only coins of values p and q, and similarly for (p, r) and (q, r), it is clear that we can pay any charge that exceeds

$$\min\{(p-1)(q-1), (q-1)(r-1), (r-1)(p-1)\}.$$

However, unlike the case of two values, *there is no known explicit expression for the maximum charge that cannot be paid*. It is known that this maximum value is *not* a polynomial in p, q and r (see [9]), so this may well be a very difficult problem.

Finally, we can solve specific problems of this type by using computer software that can multiply polynomials. Consider, for example, the problem of paying a charge N with coins of values p, q and r. Obviously, we need at most N coins of each type, so let

$$\left(\sum_{i=0}^{N} x^{pi}\right) \left(\sum_{j=0}^{N} x^{qj}\right) \left(\sum_{k=0}^{N} x^{rk}\right) = 1 + c_1 x + c_2 x^2 + \cdots.$$

As c_N is the number of terms in the expanded product that are of the form $x^{pi+qj+rk}$, where i, j and k are non-negative integers such that $pi + qj + rk = N$, we see that there are exactly c_N ways that we can pay a charge N.

2

Lengths and angles

2.1 The problem

Consider the rectangle (of height 1 and width 3) illustrated in Figure 2.1 with lengths a, b and c, and angles A, B and C.

Is there anything interesting that can be said about this figure? If so, what is it, and can it be generalised? The answer, of course, depends on what we regard as 'interesting'. However, the recognition of what is interesting, and what is not, is part of the development of a person as a mathematician. For example, although $a = \sqrt{2}$ is true, this is not very interesting because it follows trivially from Pythagoras' theorem. Likewise, we would not say that $b = \sqrt{5}$ and $c = \sqrt{10}$ are interesting. Neither is the fact that $\sqrt{2}\sqrt{5} = \sqrt{10}$, because this uses only the elementary properties of the square root function, and it is true without any reference to Figure 2.1. However, the (true) relation

$$ab = c \tag{2.1}$$

is interesting, since it immediately raises questions such as 'Will this still hold if I change the positions of the points on the base of the rectangle, and if not, does some other relationship then hold?' A fact is certainly interesting if it suggests further avenues to explore, or relationships which may hold in similar circumstances. Usually, 'interesting' statements will involve parameters.

If there are any interesting facts about Figure 2.1, then they are surely about lengths and angles, or both. We have discussed lengths, so let us now look for a relationship between the angles A, B and C. Since

$$\tan A = 1, \quad \tan B = \tfrac{1}{2}, \quad \tan C = \tfrac{1}{3}$$

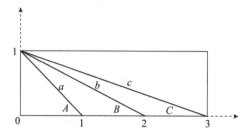

Figure 2.1 The first rectangle

(again, these are obvious and not interesting), we can use a computer
to show that *approximately* $A = 45°$, $B = 26.56°$ and $C = 18.43°$.
This suggests that

$$A = B + C, \tag{2.2}$$

but as the computed values are not exact, we have not yet proved this.
Let us now *prove* that $A = B + C$. First,

$$\tan(B + C) = \frac{\tan B + \tan C}{1 - \tan B \tan C} = \frac{\frac{1}{2} + \frac{1}{3}}{1 - \frac{1}{6}} = 1 = \tan A.$$

As $x \mapsto \tan x$ is a strictly increasing function on $(0, \pi/2)$, and negative
when $\pi/2 < x < \pi$, we see that (2.2) follows.

Now that we have proved (2.1) and (2.2) we should immediately
look for other ways to prove them. The point here is that we are not
just looking for results; instead, we are looking for *a complete understanding of the situation*, and different methods of proof might (and
usually do) give us additional insights or a different perspective. In
fact, there are many other ways of proving (2.1) and (2.2); for example, we can use complex numbers. If we draw the lines from the origin
to the points $1 + i$, $2 + i$ and $3 + i$, we easily see that $|a| = |1 + i|$,
$b = |2 + i|$ and $c = |3 + i|$. Since $(1 + i)(2 + i) = 1 + 3i$, this shows
that $ab = c$. Next, since $(2 + i)(3 + i) = 5(1 + i)$, we see that

$$\arg(2 + i) + \arg(3 + i) = \arg 5(1 + i) = \arg(1 + i).$$

It is easy to check that $\arg(1+i) = A$, $\arg(2+i) = B$ and $\arg(3+i) =
C$, so that $A = B + C$. We shall use these ideas later.

Problem 2.1   Can you find other proofs of (2.1) and (2.2)?

2.2 A generalisation

Now consider the more general situation illustrated in Figure 2.2. As before, $A = 45°$ and $a = \sqrt{2}$, so this figure is completely determined by any one of the pairs (u, v), (B, C) or (b, c).

Problem 2.2 Which constraints on u and v will give (2.1), or (2.2), or both?

If we use complex numbers in the same way as above, we see that $ab = c$ if and only if $|(1 + i)(u + i)| = |v + i|$; thus

- $ab = c$ if and only if $v^2 = 2u^2 + 1$, and
- $A = B + C$ if and only if $(u - 1)(v - 1) = 2$.

We want to look at these results in a geometric way, so it is natural to define

$$H_1 = \{(x, y) : y^2 = 2x^2 + 1\},$$
$$H_2 = \{(x, y) : (x - 1)(y - 1) = 2\}.$$

Of course, in our problem, a, b and c are positive, but for the time being, we shall ignore this (since something of interest may emerge from the more general case). Now each H_j is a hyperbola, and the reader should pause to make a rough sketch of these (note that a rough sketch is sufficient; do *not* waste time getting an accurate diagram – it will not tell you anything more than a rough sketch will). The points of intersection of H_1 and H_2 are easily found; indeed, it is clear from a geometric point of view that the two hyperbolae intersect in the first quadrant at only one point, and a little calculation shows that this is the point $(2, 3)$. From this we see that *both* (2.1) *and* (2.2) *hold if and only*

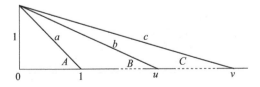

Figure 2.2 The first generalisation

if $(u, v) = (2, 3)$. As $(u, v) = (2, 3)$ corresponds to the situation in Figure 2.1, we have shown that this situation is indeed special.

Let us now confirm this result algebraically. Since $v^2 = 1 + 2u^2$ and $(u - 1)(v - 1) = 2$, the second equation gives $v = (u + 1)/(u - 1)$, and substitution in the first equation gives $4u = 2u^2(u - 1)^2$. If (u, v) is a point of intersection, then $u \neq 0$ so that $2 = u(u - 1)^2$ or $(u - 2)(u^2 + 1) = 0$. Since u is real, $u = 2$ is the only solution.

Next, we note that the points (u, v) on H_1 with $1 < u < v$ can be put in parametric form

$$(u, v) = \left(\frac{\sinh t}{\sqrt{2}}, \cosh t \right),$$

while points on H_2 have a rational parametric form, namely

$$(u, v) = \left(1 + t, 1 + \frac{2}{t} \right).$$

This may or may not be useful; nevertheless, we should *always* collect *any* information that may possibly give us some extra understanding at some later stage.

2.3 Another generalisation

A further generalisation is illustrated in Figure 2.3, where we may regard the variables m, p and q as being either positive numbers or positive integers. Again, we seek the relationships between these variables for either (2.1) or (2.2), or both, to hold.

If we follow the same arguments as above, we find that $A = B + C$ if and only if $pq = 1 + m(p + q)$ or, equivalently,

$$(p - m)(q - m) = m^2 + 1. \tag{2.3}$$

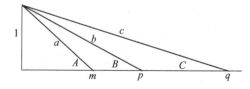

Figure 2.3 The second generalisation

Thus if we restrict m, p and q to be integers, we can give a complete solution to the problem of when $A = B + C$. We simply find all ways of writing $m^2 + 1$ as a product of two positive integers s and t, where $s < t$, and let $p = m + s$ and $q = m + t$. In each of these cases, we will have $A + B = C$. Note that there are infinitely many integer solutions of this equation in the variables m, p and q, for we can take $m = 2k + 1$, $p = m + 2$ and $q = m + 2k^2 + 2k + 1$ with $k = 1, 2, \ldots$.

Next, we find that $ab = c$ if and only if $(1+m^2)(1+p^2) = 1+q^2$; thus we now seek all positive integer solutions of

$$(1 + m^2)(1 + p^2) = 1 + q^2. \tag{2.4}$$

We note that (2.3) has solutions that are not solutions of (2.4), and (2.4) has solutions that are not solutions of (2.3): consider, for example, when (m, p, q) is $(3, 5, 8)$ or $(1, 12, 17)$. Nevertheless, equations (2.3) and (2.4) do have *infinitely many common solutions*.

Theorem 2.1 *The triples* (m, p, q), *where*

$$(m, p, q) = (n, n + 1, n^2 + n + 1), \quad n = 1, 2, \ldots,$$

are solutions of both (2.3) *and* (2.4), *and these are the only common integral solutions of these two equations.*

Proof The given triples obviously satisfy (2.3), and they satisfy (2.4) precisely because of the identity

$$(1 + x^2)\big(1 + (x + 1)^2\big) = 1 + (x^2 + x + 1)^2.$$

It follows that in the situation in Figure 2.4, for all n we have $ab = c$ and $A = B + C$ (and the case $n = 1$ is the original situation).

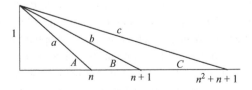

Figure 2.4 Infinitely many common solutions

Let us now show that these are the only solutions to both (2.3) and
(2.4). Suppose that (m, p, q), where $m < p < q$, is a solution of both
of these equations. If we eliminate q from these equations, we get

$$(1 + m^2)(1 + p^2) = 1 + \left(\frac{mp + 1}{p - m}\right)^2$$

or

$$(p - m)^2(1 + m^2)(1 + p^2) = (p - m)^2 + (mp + 1)^2.$$

Now the right-hand side here is $(1 + m^2)(1 + p^2)$, which is never zero,
so $(p - m)^2 = 1$. This gives $p = m \pm 1$, and since $p > m$ we have
$p = m + 1$, and hence $q = m^2 + m + 1$. We have now shown that the
triples

$$(1, 2, 3), \quad (2, 3, 7), \quad (3, 4, 13), \quad (4, 5, 22), \ldots$$

all give rise to figures that satisfy (2.3) and (2.4), and these are the only
triples of integers with this property.   □

2.4 Diophantine equations

Figure 2.1 is the special case of Figure 2.2 with $u = 2$ and $v = 3$, so
let us re-examine Figure 2.2 *with the added restriction that u and v are
now required to be positive integers*. We are now asking for all *positive
integer solutions* of each of the equations

$$Y^2 = 1 + 2X^2, \qquad (Y - 1)(X - 1) = 2,$$

and it is obvious that the only positive integral solutions (X, Y) of
$(Y - 1)(X - 1) = 2$ are $(2, 3)$ and $(3, 2)$.

Any equation of the form $P(X, Y) = 0$, where P is a polynomial
with integer coefficients, and X and Y are *integers*, is said to be a *Diophantine equation* in honour of Diophantus of Alexandria who, in the
third century, was the first to study such equations. From a geometric
point of view, a *lattice point* is a point (x, y) in the plane $\mathbb{R}^2$ where x
and y are integers, so the problem of solving a Diophantine equation is
equivalent to asking for all lattice points on the curve $P(X, Y) = 0$. In
general, this is a very difficult problem, but in our case each of the two
Diophantine equations represent a hyperbola. For a comprehensive text
on Diophantine equations, see [10].

Let us now discuss the integer solutions of $Y^2 - 2X^2 = 1$. This is a special case of a famous equation known as *Pell's equation*, the solution of which was mistakenly attributed to John Pell by none other than the famous mathematician Euler! There is a huge amount of literature on Pell's equation (see, for example, [2, 5]) which, in its most general form, is

$$Y^2 - NX^2 = 1, \qquad (2.5)$$

where N is a positive integer. The theory is sufficiently well developed to tell us how to find all integer solutions of (2.5), and the reader should certainly examine the special case when $N = M^2$ for some integer M. One possibility, then, is to put this problem aside and go and learn about Pell's equation. Note that this should not be regarded as an 'additional' task; instead, the original problem should be regarded as a catalyst that has led us to some worthwhile mathematics to learn. We repeat, the solution of the original problem is *not* the ultimate goal; the role of the problem is primarily to stimulate interest in, and the understanding of, more mathematics.

Let us continue to explore the integer solutions of (2.5), but without assuming that the reader has any knowledge of Pell's equation. Clearly, if $N = 2$, then $(X, Y) = (2, 3)$ is a solution, but we do not yet know whether there are any other integer solutions. If there are, how many solutions are there? Are there infinitely many solutions, or only a finite number of solutions? Can we write down *all* integer solutions?

First, we can find some integer solutions of $Y^2 - 2X^2 = 1$ by trial and error (that is, by searching on a computer), and if we do this, we obtain the solutions

$$(0, 1), (2, 3), (12, 17), (70, 99), (408, 577), (2378, 3363), \dots.$$

The solutions appear to be growing rapidly, and no obvious general formula presents itself. Perhaps it is now time to look at texts on number theory to obtain more information about Pell's equation? Pell's equation has been studied for over 2,500 years (although not, of course, with Pell's name attached), with valuable contributions from the early Greek and Indian mathematicians. For example, in about 600 AD the Indian mathematician Brahmagupta knew that we can compose two solutions to obtain a third solution: for example, if $Y^2 - 2X^2 = 1$ and $V^2 - 2U^2 = 1$ then

$$(YV + 2XU)^2 - 2(XV + YU)^2 = (Y^2 - 2X^2)(V^2 - 2U^2) = 1,$$

so that we have a third solution $(YV + 2XU, XV + YU)$. In modern terminology, this can be expressed in terms of matrices. Given a solution (X_1, Y_1), so that $Y_1^2 - 2X_1^2 = 1$, we define the matrix

$$M_1 = \begin{pmatrix} Y_1 & 2X_1 \\ X_1 & Y_1 \end{pmatrix}. \tag{2.6}$$

Note that $\det(M_1) = 1$. If (X_2, Y_2) is another solution, and M_2 is the corresponding matrix, then Brahmagupta's observation is simply that $\det(M_1 M_2) = \det(M_1)\det(M_2) = 1$. It now follows that each matrix M_1^n, $n = 1, 2, 3, \ldots$ generates a solution of $X^2 - 2Y^2 = 1$. Thus, by taking $(X_1, Y_1) = (2, 3)$, and *multiplying matrices on a computer*, we can generate infinitely many integral solutions of the equation $Y^2 - 2X^2 = 1$. An obvious question now is: *Does this method produce all positive integral solutions of* $Y^2 - 2Y^2 = 1$? The answer is 'yes', and for more information the reader should now visit other sources of information on Pell's equation.

The reader should now check that the method described above does produce the solutions given above, and that the next solution (obtained from M_1^6) is (13860, 19601); thus $19601^2 - 2 \times 13860^2 = 1$. Notice that this means that $(19601/13860)^2$ is almost 2, so that $19601/13860$ must be a good approximation to $\sqrt{2}$. *How close is this to* $\sqrt{2}$? In general, large solutions of (2.5) yield good approximations to $\sqrt{N}$.

2.5 The value of π

Early estimates of π were found by constructing both inscribed and circumscribed regular polygons about the unit circle, and then finding the lengths of these polygons. Indeed, Archimedes used this method to show that π lies between $3\frac{10}{71}$ and $3\frac{1}{7}$. This latter value, namely $\frac{22}{7}$, is a commonly used approximation, and a better approximation, namely $\frac{355}{113}$, was found by Tsu Ch'ung-chih in about 470 AD. Much later, Ludolph van Ceulen (1539–1610) obtained π to 35 decimal places.

In 1671, Gregory found the power series for the inverse tangent function, namely

$$\tan^{-1} x = x - \frac{x^3}{3} + \frac{x^5}{5} - \cdots,$$

and this implies that

$$\frac{\pi}{4} = 1 - \frac{1}{3} + \frac{1}{5} - \cdots,$$

which was discovered independently by Leibnitz in 1673. This series converges far too slowly to be of much use in estimating π (the reader should estimate how many terms need to be computed to obtain π to, say, six decimal places). However, by using the formula

$$\tan^{-1} x + \tan^{-1} y = \tan^{-1} \left(\frac{x+y}{1-xy} \right),$$

(valid for positive x and y) we can express π as a linear combination of terms of the form $\tan^{-1} x$, which can then be computed more efficiently because, when $0 < x < 1$, the series for $\tan^{-1} x$ converges at a faster rate. The most well known of these is Machin's formula

$$\frac{\pi}{4} = 4\tan^{-1}\frac{1}{5} - \tan^{-1}\frac{1}{239},$$

which follows directly from the fact that $(5+i)^4 = (2+2i)(239+i)$. In about 1700, Machin found π to 100 decimal places, and in 1853 Shanks determined it to 607 decimal places. From a more theoretical viewpoint, Lambert proved (around 1700) that π is irrational, and in 1882 Lindemann proved that it is transcendental (that is, it is not the root of any polynomial with integral coefficients).

At this point, the reader may well be wondering what has all this to do with the earlier work? Well, the fact that $A = B + C$ in Figure 2.1 is equivalent to the assertion that

$$\frac{\pi}{4} = \tan^{-1}\frac{1}{2} + \tan^{-1}\frac{1}{3},$$

a formula obtained long ago by Euler. More generally, Theorem 2.1 implies that, for each positive integer n,

$$\tan^{-1}\frac{1}{n} = \tan^{-1}\frac{1}{n+1} + \tan^{-1}\frac{1}{1+n+n^2};$$

for example,

$$\tan^{-1}\frac{1}{2} = \tan^{-1}\frac{1}{3} + \tan^{-1}\frac{1}{7}.$$

From these formulae, we can obtain many expressions for $\pi/4$ as a linear combination of terms of the form $\tan^{-1} x$; for example (as the reader can, and should, check), if $0 < b < a$ then

$$\tan^{-1} \frac{b}{a} + \tan^{-1} \frac{a-b}{a+b} = \frac{\pi}{4}.$$

We leave the reader to explore these ideas further; however, for more details, see [4].

3
Magic squares

3.1 Magic squares

Magic squares were known to the Chinese over 2,600 years ago, and a Chinese legend tells of a turtle emerging from the sea with the 3×3 magic square

$$\begin{pmatrix} 4 & 9 & 2 \\ 3 & 5 & 7 \\ 8 & 1 & 6 \end{pmatrix}$$

engraved on its shell. So what is a magic square? A real $n \times n$ matrix X is a *magic square* if the sums of its elements taken over each row, over each column and over each of the two diagonals all have the same value. Thus if $X = (x_{i,j})$, then X is a magic square if and only if, for $i, j = 1, \ldots, n$; likewise for some K we have

$$K = x_{i,1} + x_{i,2} + \cdots + x_{i,n}, \tag{3.1}$$

$$K = x_{1,j} + x_{2,j} + \cdots + x_{n,j}, \tag{3.2}$$

$$K = x_{1,1} + \cdots + x_{n,n}, \tag{3.3}$$

$$K = x_{1,n} + x_{2,(n-1)} + \cdots x_{n,1}. \tag{3.4}$$

We say that X is a *semi-magic square* if (3.1) and (3.2) (the sums over the rows and the columns), but not necessarily (3.3) or (3.4) (the sums over the two diagonals), hold. In each case, we call K the *magic constant* of X and denote it by $\mu(X)$.

Problem 3.1 Each 1×1 matrix is a magic square. Find the most general 2×2 magic square. Find the most general 2×2 semi-magic square.

Problem 3.2   Check that

$$\begin{pmatrix} 7 & 12 & 1 & 14 \\ 2 & 13 & 8 & 11 \\ 16 & 3 & 10 & 5 \\ 9 & 6 & 15 & 4 \end{pmatrix} \tag{3.5}$$

is a magic square whose entries are $1, 2, \ldots, 16$. This magic square was engraved during the tenth century in stone in the Parshvanath Jain temple in India.

Problem 3.3   Let E_n be the $n \times n$ matrix with all entries equal to 1. Show that E_n is a magic square, $\mu(E_n) = n$ and aE_n is a magic square with $\mu(aE_n) = na$.

Problem 3.4   Suppose that X is an $n \times n$ semi-magic (or magic) square. Show that its magic constant $\mu(X)$ of X is n times the average value of the coefficients of X.

Problem 3.5   Construct a semi-magic square that is not a magic square.

We want to investigate magic squares, and semi-magic squares, from a mathematical perspective and explore some of their interesting properties. Some people insist that the coefficients of an $n \times n$ magic square are precisely the integers $1, 2, \ldots, n^2$, and others take a more relaxed view and only require that the coefficients are integers. However, although these restrictions offer a serious challenge, there is more interesting mathematics to be found without them. We shall allow the entries *to be any real numbers* for then we have the obvious, but mathematically significant, result that any linear combination of $n \times n$ magic squares is again a magic square. This means that the set $\mathcal{M}^n$ of real $n \times n$ magic squares *is a real vector space*, and the same is true for the set $\mathcal{S}^n$ of real $n \times n$ semi-magic squares. As each of these spaces is a subspace of the vector space of all real $n \times n$ matrices (which has dimension n^2), we see that each has dimension at most n^2. We must surely now find the dimensions of these spaces, and we shall discuss this later.

Problem 3.6   Suppose that X and Y are $n \times n$ semi-magic squares. Is it true that $\mu(XY) = \mu(X)\mu(Y)$? Does this equation hold if X and

Y are magic squares? You may wish to try some numerical examples
before trying to answer these questions.

3.2 The space of 3 × 3 magic squares

We begin with an elementary, but useful, observation. Let $X = (x_{i,j})$
be a 3×3 magic square with magic constant K. If we sum the elements
in the middle row, the middle column and each diagonal, and add these
together, we obtain $4K$. On the other hand, the answer is obviously
$\sum_{i,j} x_{ij} + 3x_{22}$, which is $3K + 3x_{22}$. We deduce that the 'central'
coefficient x_{22} is $\mu(X)/3$.

Now let X be a 3×3 magic square with magic constant K. Then
$X - (K/3)E_3$ (recall that E_3 has all entries equal to 1) is a magic square
with magic constant 0; thus, from now on *we can confine our attention
to* 3×3 *magic squares with magic constant* 0. Now let X be a 3×3
magic square with magic constant 0, say

$$X = \begin{pmatrix} a & * & b \\ * & 0 & * \\ * & * & * \end{pmatrix},$$

where the entries designated by $*$ are as yet unknown. As $\mu(X) = 0$,
it follows immediately that

$$X = \begin{pmatrix} a & -a-b & b \\ b-a & 0 & a-b \\ -b & a+b & -a \end{pmatrix}, \tag{3.6}$$

so this is the most general 3×3 magic square with magic constant 0. It
is convenient to write this as $X = aA + bB$, where the magic squares
A and B, each with magic constant 0, are given by

$$A = \begin{pmatrix} 1 & -1 & 0 \\ -1 & 0 & 1 \\ 0 & 1 & -1 \end{pmatrix}, \quad B = \begin{pmatrix} 0 & -1 & 1 \\ 1 & 0 & -1 \\ -1 & 1 & 0 \end{pmatrix}. \tag{3.7}$$

Problem 3.7 Show that $\{E_3, A, B\}$ is a basis of $\mathcal{M}^3$, so that
$\dim(\mathcal{M}^3) = 3$. Show further that $\{A, E_3\}$ is a basis of the space of
symmetric 3×3 magic squares, and that $\{B\}$ is a basis of the space of
skew-symmetric 3×3 magic squares.

We shall now establish a result about the geometric action of a 3×3 magic square X. It is convenient to regard a point v in $\mathbb{R}^3$ as a column vector; then X acts on $\mathbb{R}^3$ as the map $v \mapsto Xv$, where this is computed by matrix multiplication. Now let

$$u = \begin{pmatrix} 1 \\ 1 \\ 1 \end{pmatrix}, \quad v = \begin{pmatrix} 1 \\ 0 \\ -1 \end{pmatrix}, \quad w = \begin{pmatrix} 1 \\ -2 \\ 1 \end{pmatrix}.$$

The reader should check that these three vectors are mutually orthogonal to each other, and that if Π is the plane given by $x + y + z = 0$, then v and w are in Π, and u is the normal to Π. We let U, V and W be the lines (or one-dimensional subspaces) spanned by u, v and w, respectively.

Theorem 3.1 *Let X be a 3×3 real matrix. Then X is a magic square if and only if*

$$X(U) \subset U, \quad X(V) \subset W, \quad X(W) \subset V. \tag{3.8}$$

Proof Let A, B and E_3 be as above. It is easy to check that each of these three matrices maps (i) u into U, (ii) v into W and (iii) w into V. As these three matrices span $\mathcal{M}^3$, any magic square X is a linear combination of them, so that X satisfies (3.8).

Now suppose that X is a 3×3 matrix that satisfies (3.8); then it is sufficient to show that X is a linear combination of E_3, A and B. Now, from (3.8), there are scalars a, b and c such that $Xu = au$, $Xv = bw$ and $Xw = cv$. We now leave it to the reader to show that for a suitable choice of α, β and γ, if $Y = \alpha A + \beta B + \gamma E_3$, then $X = Y$ on $\{u, v, w\}$. Then, as $\{u, v, w\}$ is a basis of $\mathbb{R}^3$, we see that $X = Y$. This completes the proof. □

The next result follows from Theorem 3.1.

Corollary 3.2 *Suppose that X is a 3×3 magic square. Then every odd power of X is a magic square, and if X is a non-singular then X^{-1} is also magic square.*

Proof The proof of the first statement follows from (3.8). Now assume that X is non-singular. Then, by Theorem 3.1, $X(U) \subset U$,

$X(V) \subset W$ and $X(W) \subset V$. As X is non-singular, $X(U)$, $X(V)$ and $X(W)$ each have dimension one so that $X(U) = U$, $X(V) = W$ and $X(W) = V$. Therefore, $X^{-1}(U) = U$, $X^{-1}(V) = W$ and $X^{-1}(W) = V$ so that, by Theorem 3.1 again, X^{-1} is a magic square.   □

Problem 3.8   When is the magic square (3.6) non-singular?

Corollary 3.2 raises the question of whether or not an even power of a 3×3 magic square is a magic square, and we ask the reader to provide an example now to show that it need not be. This matter is discussed further in the next problem.

Problem 3.9   Let X be a 3×3 magic square with $\mu(X) = 0$. Show that we can write $X = cC + dD$ for some real numbers c and d, where $C = A + B$ and $D = A - B$, and A and B are as given in (3.7). Now compute the matrices C^2, D^2, $(CD)^2$ and $(DC)^2$, and use these to simplify the expression $(cC + dD)^n$ (when expanded) for X^n. Now comment on whether or not an even power of a 3×3 magic square is a magic square.

3.3 The space of 4×4 magic squares

We begin with two problems.

Problem 3.10   Show that N, where

$$
N = \begin{pmatrix} 15 & -11 & -13 & 9 \\ -7 & 3 & 5 & -1 \\ 1 & -5 & -3 & 7 \\ -9 & 13 & 11 & -15 \end{pmatrix}, \tag{3.9}
$$

is a magic square. What is $\mu(N)$?

Problem 3.11   Show that for all real numbers a, b, c, d, A, B, C and D the matrix M given by

$$M = \begin{pmatrix} A-a & C+a+c & B+b-c & D-b \\ D+a-d & B & C & A-a+d \\ C-b+d & A & D & B+b-d \\ B+b & D-a-c & A-b+c & C+a \end{pmatrix} \quad (3.10)$$

is a magic square. This, which was first found by C. Bergholt in 1910, is the most general 4×4 magic square. What is $\mu(M)$? Prove (formally) that the space of such matrices has dimension eight. Check that the matrices in (3.5) and (3.9) are of this form.

The reader should now recall the Cayley–Hamilton theorem that *every matrix satisfies its own characteristic equation* (if the reader has not met this result before, now is the time to learn about it!).

Problem 3.12 Show that N, given in (3.9), has characteristic equation $x^4 - 256x^2 = 0$ so that, by the Cayley–Hamilton theorem, $N^4 = 256N^2$. The reader should now check (directly, on a computer) that although $N^4 = 256N^2$ and $N^3 = 256N$, it is *not* true that $N^2 \neq 256I$, where I is the 4×4 identity matrix. Deduce that (i) N does not have an inverse and (ii) N^p is a magic square for every positive, odd integer p.

3.4 The dimension of the space of magic squares

We now consider the problem of finding the dimensions of the spaces $\mathcal{M}^n$ and $\mathcal{S}^n$ of $n \times n$ magic squares and semi-magic squares, respectively, where $n \geqslant 4$. To obtain some insight into this problem we observe that if X is an $n \times n$ magic square, then it has n^2 coefficients that are subject to $2n + 1$ linear constraints (obtained by equating the sum over the first row to the sums over the other $n - 1$ rows, to the n columns and to the two diagonals). These linear constraints reduce to, say, q linearly independent constraints, and then (from the general theory of linear equations) we find that $\dim(\mathcal{M}^n) = n^2 - q$. As $g \leqslant 2n + 1$, this gives $\dim(\mathcal{M}^n) \geqslant n^2 - 2n - 1$, and this (for large n) gives us a plentiful supply of $n \times n$ magic squares.

Problem 3.13 Use the fact that the sum of the 'row-sums' is equal to the sum of the 'column-sums' to show that $q \leqslant 2n$. Thus $\dim(\mathcal{M}^n) \geqslant n(n-2)$ (and this is, in fact, an equality).

Let $\mathcal{S}^n$ be the vector space of all $n \times n$ semi-magic squares, and let $\mathcal{S}_0^n$ be the subspace of semi-magic squares M with magic constant 0. We shall now sketch an argument for the case $n = 3$, *but its main purpose is to guide the reader through the proof in the case of a general* n. The map

$$\Theta : \begin{pmatrix} a & b \\ c & d \end{pmatrix} \mapsto \begin{pmatrix} a & b & -(a+b) \\ c & d & -(c+d) \\ -(a+c) & -(b+d) & a+b+c+d \end{pmatrix} \quad (3.11)$$

shows that any 2×2 real matrix extends uniquely to a 3×3 semi-magic square with magic constant 0, and it is clear that every 3×3 semi-magic square with magic constant 0 arises in this way. As Θ is an isomorphism from the vector space of 2×2 real matrices onto $\mathcal{S}_0^3$, this shows (when $n = 3$) that $\dim(\mathcal{S}_0^n) = (n-1)^2$.

Problem 3.14 Show that $\dim(\mathcal{S}_0^n) = (n-1)^2$. Use (i) that $\mathcal{S}_0^n$ is the kernel of the linear map μ that takes a semi-magic square to its magic constant, (ii) μ maps *onto* $\mathbb{R}$ and (iii) the rank-kernel theorem, to show that $\dim(\mathcal{S}^n) = (n-1)^2 + 1$.

The next result gives a geometric characterisation of $n \times n$ semi-magic squares with magic constant 0. We regard $\mathbb{R}^n$ as the space of column vectors, and we let

$$\mathbf{e}_1 = \begin{pmatrix} 1 \\ 0 \\ \vdots \\ 0 \end{pmatrix}, \cdots, \mathbf{e}_n = \begin{pmatrix} 0 \\ 0 \\ \vdots \\ 1 \end{pmatrix}, \quad \mathbf{e} = \begin{pmatrix} 1 \\ 1 \\ \vdots \\ 1 \end{pmatrix}.$$

The subspace generated by a collection $\{\mathbf{u}_1, \ldots\}$ of vectors in $\mathbb{R}^n$ is denoted by $\langle \mathbf{u}_1, \ldots \rangle$; in particular, we let

$$E = \{t\mathbf{e} : t \in \mathbb{R}\} = \langle \mathbf{e} \rangle,$$
$$\Pi = \{(x_1, \ldots, x_n)^t : x_1 + \cdots + x_n = 0\} = \{\mathbf{x} : \mathbf{x} \cdot \mathbf{e} = 0\}$$

of $\mathbb{R}^n$. The reader should consider the geometry here in the case $n = 3$. Clearly, $\dim(E) = 1$, $\dim(\Pi) = n - 1$ and E is the orthogonal complement of Π. Each $n \times n$ matrix M acts on $\mathbb{R}^n$ by matrix multiplication, namely $\mathbf{x} \mapsto M\mathbf{x}$, and we denote this linear transformation by α_M.

Problem 3.15 Prove that an $n \times n$ matrix M is a semi-magic square with magic constant 0 if and only if $E \subset \ker(\alpha_M)$ and $\alpha_M(\Pi) \subset \Pi$.

The result in this problem provides a proof that any finite product of semi-magic squares with magic constant 0 is again a semi-magic square with magic constant 0. It also shows that a semi-magic square with magic constant 0 is, in effect, a linear transformation of Π into itself and, as $\dim(\Pi) = n - 1$, this provides an alternative proof that $\dim(\mathcal{S}_0^n) = (n - 1)^2$.

3.5 Lagrange's interpolation formula

First, we consider Lagrange's formula for a real variable. Given n distinct real numbers $x_1, \ldots, x_n$, and n real numbers $y_1, \ldots, y_n$, there is a unique polynomial f of degree $n - 1$ such that $f(x_j) = y_j$ for $j = 1, \ldots, n$, and f is given explicitly by *Lagrange's interpolation formula*. To express Lagange's result succinctly, we shall use the notation $\prod_{j \setminus i} x_j$ to denote the product of the x_j for $j \in \{1, 2, \ldots, n\} \setminus \{i\}$ (that is, for $j = 1, \ldots, n$ but excluding $j = i$); then Lagrange's formula for the polynomial f is

$$f(x) = \sum_{i=1}^{n} \left(\prod_{j \setminus i} \frac{(x - x_j)}{(x_i - x_j)} \right) y_i.$$

It does not seem to be well known that Lagrange's interpolation formula also holds for a square matrix, and it is clear that this will be useful in examining whether, given a magic square X and a polynomial f, the matrix $f(X)$ is also a magic square. Here we shall focus on Lagrange's formula for matrices and leave the reader to explore the consequences of this for magic squares.

Lagrange's formula for matrices depends heavily on the eigenvalues of the given matrix, and, for simplicity, we shall only discuss the case where the eigenvalues are distinct. Let M be an $n \times n$ magic matrix with distinct eigenvalues $\lambda_1, \ldots, \lambda_n$, and characteristic equation $\chi_M(x) = 0$; thus $\chi_M(x) = (x - \lambda_1) \cdots (x - \lambda_n)$. Suppose now that f and g are two polynomials such that

$$f(\lambda_i) = g(\lambda_i), \quad 1 = 1, \ldots, n,$$

and let $h = f - g$. Then $h(\lambda_i) = 0$ for each i, so that χ_M divides h. Since M satisfies its own characteristic equation, we see that $h(M) = \mathcal{O}$; hence $f(M) = g(M)$. Thus $f(M)$ *is uniquely determined by the values* $f(\lambda_1), \ldots, f(\lambda_n)$. It follows immediately that if M has distinct eigenvalues $\lambda_1, \ldots, \lambda_n$, and f is any polynomial, then

$$f(M) = \sum_{i=1}^{n} \left(\prod_{j \backslash i} \frac{(M - \lambda_j \mathcal{I})}{(\lambda_i - \lambda_j)} \right) f(\lambda_i);$$

this is simply because the two polynomials

$$f(x), \quad \sum_{i=1}^{n} \left(\prod_{j \backslash i} \frac{(x - \lambda_j)}{(\lambda_i - \lambda_j)} \right) f(\lambda_i)$$

have the same value at each λ_i. In particular, for any positive integer p, we have *Lagrange's interpolation formula for matrices*:

$$M^p = \sum_{i=1}^{n} \left(\prod_{j \backslash i} \frac{(M - \lambda_j \mathcal{I})}{(\lambda_i - \lambda_j)} \right) \lambda_i^p. \tag{3.12}$$

This discussion shows that once we know the eigenvalues of M, and also the matrices $M^2, \ldots, M^{n-1}$, we can, in principle, determine whether any matrix $f(M)$ is magic or not.

Problem 3.16 Explore Lagrange's interpolation formula when applied to 2×2 and 3×3 matrices.

We end with a completely open-ended problem.

Problem 3.17 Show that the matrix

$$\begin{pmatrix} 23 & 1 & 2 & 20 & 19 \\ 22 & 16 & 9 & 14 & 4 \\ 5 & 11 & 13 & 15 & 21 \\ 8 & 12 & 17 & 10 & 18 \\ 7 & 25 & 24 & 6 & 3 \end{pmatrix}$$

is a magic square with magic constant 65. Use a computer algebra package to explore the properties of powers of this matrix; for example, can you find any powers that are also magic squares?

4

Intersecting chords

4.1 Dividing a disc by chords

First, some notation. We denote the complex plane by $\mathbb{C}$, the unit circle $\{z : |z| = 1\}$ by $\mathcal{C}$, and its interior $\{z : |z| < 1\}$ by $\mathbb{D}$. We use $[a, b]$ for the line segment in $\mathbb{C}$ with endpoints a and b.

Now let n be an integer, with $n \geqslant 2$. Choose n distinct points $z_1, \ldots, z_n$ on $\mathcal{C}$, and draw the chords $[z_i, z_j]$, where $i, j = 1, 2 \ldots, n$ and $i \neq j$; the case $n = 4$ is illustrated in Figure 4.1.

If $n \geqslant 5$ it may happen that three (or more) chords intersect at some point strictly inside $\mathcal{C}$; however, if this happens we can always move the z_j slightly to avoid such intersections.

Problem 4.1 Show that if $z_1, \ldots, z_n$ are given, then at most two chords constructed from $z_1, \ldots, z_n, z_{n+1}$ pass through any point of $\mathbb{D}$ providing only that the additional point z_{n+1} avoids a finite set of points on the unit circle.

Thus, from now on, we shall assume that *at most two of the chords* $[z_i, z_j]$ *pass through any point of* $\mathbb{D}$. Since there are n points z_i, there are $\binom{n}{2}$ chords $[z_i, z_j]$, and these divide $\mathbb{D}$ into a certain number of regions. Let the number of regions be R_n; for example, if $n = 4$ then there are six chords and eight regions (so that $R_4 = 8$).

Problem 4.2 What are R_2, R_3, R_4 and R_5? Does R_n depend only on n and *not on the positions of the* z_j? If so, can you find R_n as an explicit function of n?

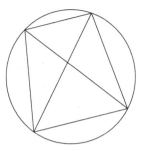

Figure 4.1 The case $n = 4$

The reader should now verify (by drawing pictures) that $R_2 = 2$, $R_3 = 4$, $R_4 = 8$ and $R_5 = 16$. Perhaps $R_n = 2^{n-1}$ for all n? If the reader has indeed checked these cases it will be clear that it would not be sensible to try to find, say, R_{100} by drawing 4,950 chords, and then counting the regions. Indeed, if the answer suggested above is correct (and it may not be), we would have 2^{4949} regions to count! How, then, can we find R_{100}?

4.2 The recurrence relation for R_n

Suppose that the points $z_1, \ldots, z_n$ are given, and that the $\binom{n}{2}$ chords divide $\mathbb{D}$ into R_n regions. We are now going to count the number of *additional* regions that we obtain by adding the additional point z_{n+1}, and then drawing each of the n additional chords $[z_1, z_{n+1}], \ldots, [z_n, z_{n+1}]$. Let the number of additional regions be A_n. The idea is that if we can find A_n explicitly then we will have the recurrence relation $R_{n+1} = R_n + A_n$ which, we hope, we can solve.

We may suppose (by relabelling, if necessary) that the points $z_1, \ldots, z_{n+1}$ are arranged in this order around $\mathcal{C}$. As the chords $[z_1, z_{n+1}], \ldots, [z_n, z_{n+1}]$ (all from z_{n+1}) do not intersect each other in $\mathbb{D}$, the number A_n of additional regions obtained by drawing all of these chords is the same as the sum over $k = 1, \ldots, n$ of the number a_k of the additional regions obtained by only drawing the chord $[z_k, z_{n+1}]$; thus $A_n = a_1 + a_2 + \cdots + a_n$. We shall now find a_k, then A_n, and then, we hope, R_n.

Let us now choose any k in $\{1, \ldots, n\}$, and then add the chord $[z_k, z_{n+1}]$ to the picture already drawn for the given n. This chord will cross a given chord, say $[z_i, z_j]$, where $i, j \in \{1, \ldots, n\}$, if and only if z_i and z_j lie on different sides of the chord $[z_k, z_{n+1}]$. Since the points $z_1, \ldots, z_{k-1}$ lie on one side of $[z_{n+1}, z_k]$, and the points $z_{k+1}, \ldots, z_n$ lie on the other side of $[z_{n+1}, z_k]$, the number of points on $[z_{n+1}, z_k]$ where this chord meets one of the given chords is $(k-1)(n-k)$. Let us label these intersection points as $\zeta_1, \ldots, \zeta_r$, where $r = (k-1)(n-k)$ and where these points occur in this order along the chord. Then each of the segments

$$[z_{n+1}, \zeta_1], [\zeta_1, \zeta_2], \ldots, [\zeta_{r-1}, \zeta_r], [\zeta_r, z_k]$$

will cross (and subdivide) one of the regions shown in the original picture (for the given n). This shows that $a_k = r + 1$; hence

$$a_k = -k^2 + (n+1)k + (1-n),$$

and this gives us the fundamental recurrence relation

$$R_{n+1} = R_n + \sum_{k=1}^{n} [-k^2 + (n+1)k + (1-n)], \qquad (4.1)$$

with initial condition $R_2 = 2$. We can now check that $R_2, \ldots, R_5$ are as given above and that (more interestingly) $R_6 = 31$ which is *not* 2^5. Thus our earlier suggestion that R_n might be 2^{n-1} is wrong. The relation (4.1) allows us (in principle) to *compute* R_n for any n, and the reader should check, for example, that $R_9 = 163$. In fact, it is more efficient to simplify the formula for A_n before doing any calculations.

The reader should know (or verify) that

$$\sum_{k=1}^{n} k = \frac{n(n+1)}{2},$$

$$\sum_{k=1}^{n} k^2 = \frac{n(n+1)(2n+1)}{6},$$

$$\sum_{k=1}^{n} k^3 = \frac{n^2(n+1)^2}{4}.$$

If we use these with (4.1), we obtain the relation

$$R_{n+1} - R_n = \frac{n(n^2 - 3n + 8)}{6}. \qquad (4.2)$$

If we now sum both sides of this equation over $n = 2, \ldots, N - 1$, and use $R_2 = 2$, we obtain

$$R_n = 2 + \sum_{n=2}^{N-1} \frac{n(n^2 - 3n + 8)}{6},$$

which (after a little algebra) leads to an explicit formula for R_n, namely

$$R_n = \frac{n^4 - 6n^3 + 23n^2 - 18n + 24}{24}. \tag{4.3}$$

This is the solution to the given problem, but it is not the end of the discussion. The reader should check (again using a little algebra) that (4.3) gives

$$R_n = 1 + \binom{n}{2} + \binom{n}{4}. \tag{4.4}$$

This closed expression for R_n surely suggests that there might be another (and better?) way to obtain this result, and we shall now take this idea further. In any event, we can now see that $R_{100} = 3926176$.

4.3 Lagrange's interpolation formula

In this section, we present another way to continue once we have established (4.1).

Problem 4.3 Prove that, for any positive integer p, the sum $\sum_{k=1}^{n} k^p$ is a polynomial of degree $p + 1$ in the variable n. Note that the cases $p = 1, 2, 3$ are given above.

Since, for any positive integer p, the sum $\sum_{k=1}^{n} k^p$ is a polynomial of degree $p + 1$ in the variable n, it is immediately evident from (4.1) that R_n is a polynomial of degree four in n. We shall now use this to give an alternative way to find the polynomial R_n.

Lagrange's interpolation formula shows that if we take any $n + 1$ distinct points $x_1, \ldots, x_{n+1}$, and any $n + 1$ values $y_1, \ldots, y_{n+1}$ (which need not be distinct), then *there is a unique polynomial q of degree n such that $q(x_j) = y_j$ for $j = 1, 2, \ldots, n + 1$, and, moreover, the* *polynomial q can be written down explicitly.* As we know that R_n is a polynomial of degree four, and

$$R_2 = 2, \ R_3 = 4, \ R_4 = 8, \ R_5 = 16, \ R_6 = 31,$$

Lagrange's interpolation formula will give us an explicit formula for R_n without recourse to recurrence relations. We shall give a brief introduction to Lagrange's formula below (but see Chapter 3), but the reader should surely master the ideas, use them to find the formula for R_n and then, *of course*, check that the formula agrees with the formula for R_n which we already have. Indeed, the reader should, wherever possible, always check the progress made by using a different method.

Suppose that we are looking for the unique polynomial p of degree 3 such that $p(x_i) = y_i$ for $i = 1, 2, 3, 4$, where x_1, x_2, x_3, x_4 are distinct. Let

$$p_k(x) = \frac{(x - x_1)(x - x_2)(x - x_3)(x - x_4)}{x - x_k},$$

or, explicitly,

$$p_1(x) = (x - x_2)(x - x_3)(x - x_4),$$
$$p_2(x) = (x - x_1)(x - x_3)(x - x_4),$$
$$p_3(x) = (x - x_1)(x - x_2)(x - x_4),$$
$$p_4(x) = (x - x_1)(x - x_2)(x - x_3).$$

Then, obviously,

$$p_k(x_j) \begin{cases} = 0 & \text{if } j \neq k, \\ \neq 0 & \text{if } j = k. \end{cases}$$

Given this, the desired polynomial is (obviously)

$$y_1 \frac{p_1(x)}{p_1(x_1)} + y_2 \frac{p_2(x)}{p_2(x_2)} + y_3 \frac{p_3(x)}{p_3(x_3)} + y_4 \frac{p_4(x)}{p_4(x_4)},$$

since for each j, this takes the value y_j at x_j.

Problem 4.4   We know that if $p(x) = x^2 + 1$ then $p(1) = 2, p(2) = 5$ and $p(3) = 10$. Now pretend that you do not know this and use Lagrange's interpolation formula to find the quadratic polynomial q that satisfies $q(1) = 2, q(2) = 5$ and $q(3) = 10$.

4.4 Euler's formula

We shall now derive the formula (4.4) for R_n from a completely different point of view, namely from Euler's famous formula for a graph lying in the plane. We shall not state Euler's theorem in great generality, nor shall we prove it here. We do hope, however, that once the reader has seen the formula in action he or she will be tempted to find out more about it. Since the cases $n = 2$ and $n = 3$ are trivial, we shall assume that $n \geqslant 4$ (for only in these cases are there chords that intersect somewhere inside the circle).

First, we consider the unit circle divided into n arcs by the n points $z_1, \ldots, z_n$. If we 'remove' these arcs we are left with a polygon bounded by the n chords

$$[z_1, z_2], [z_2, z_3], \ldots, [z_{n-1}, z_n], [z_n, z_1],$$

and all of the other chords lie 'inside' this polygon. The points of intersection of the chords in this polygon (including the points z_i) are called *vertices*; the straight segments joining any two of these vertices (excluding the arcs of the unit circle) are called *edges*; the plane regions that are enclosed by a chain of edges are called *faces*. Euler's formula tells us that if we have F_n faces, E_n edges and V_n vertices, then

$$F_n - E_n + V_n = 1.$$

It is important to understand (without going into detail here) that *Euler's formula applies to any figure, or 'graph', with vertices, edges and faces*, and not just to the situation we are interested in here. We have applied it here to solve a particular problem but, in fact, *it is an extremely powerful and very general formula*.

Let us now use Euler's formula to find R_n. Since there are n regions that are each bounded by a single chord and a single arc of the unit circle, and there are F_n faces that are bounded only by chords, we see that $R_n = F_n + n$. Thus Euler's formula gives

$$R_n = E_n - V_n + n + 1. \tag{4.5}$$

It is obvious that given four given distinct points z_i on the circle, we obtain six chords by joining each pair of these points, and that these chords contribute just one intersection point inside the circle. Conversely, each intersection point arises in this way (and determines the set of four points z_i uniquely, since we have excluded the possibility that three or more chords meet at a single point). This shows that there are exactly $\binom{n}{4}$ vertices of the polygon inside the circle. Since there are also n vertices of the polygon on the circle, we see that

$$V_n = n + \binom{n}{4}.$$

In view of (4.4), this looks very hopeful.

We must now find E_n, and to do this we count *the ends of all edges* in two different ways (we can imagine each edge to have two outward pointing arrowheads at its end, and we shall count the number of arrowheads). First, since each of the E_n edges has two ends, this count obviously gives $2E_n$. Next, there are n vertices on the unit circle, and each is the end-point of $n - 1$ chords; this contributes $n(n - 1)$ to the count of the ends of the edges. Second, there are $\binom{n}{4}$ vertices inside the circle, and each is the endpoint of four edges. This contributes $4\binom{n}{4}$ to the count, so we obtain

$$2E_n = 4\binom{n}{4} + 2\binom{n}{2}.$$

These formulae for E_n and V_n, together with (4.5), give (4.4).

4.5 Similar problems

There are many variations of the problem we have discussed above, and we now mention a few of these. Of course, readers are encouraged to solve these problems themselves, using *all* of the techniques illustrated above. It is strongly recommended that a problem is solved *in as many ways as possible*, as each extra solution usually gives an added insight into both the problem and the solution.

Problem 4.5 Draw n distinct lines in the plane, and assume that no pair of lines are parallel and that at most two lines meet at any point. These lines divide the plane into, say R_n, regions. Find a formula for R_n.

Experiments show that $R_1 = 2$, $R_2 = 4$, $R_3 = 7$, $R_4 = 11$ and $R_5 = 16$. An argument similar to that given above shows that $R_{n+1} = n + 1 + R_n$, and the reader should find that

$$R_n = \binom{n}{2} + n + 1.$$

The ambitious reader should now try to find the formula for R_n when (i) some of the lines are parallel, or (ii) at least three lines meet at a point, or (iii) when both of these possibilities occur.

Problem 4.6 Consider distinct points $z_1, \ldots, z_n$ in the plane $\mathbb{C}$ such that no three of the z_i are collinear. Let L_j be the line that contains the points z_i and z_j, where $i \neq j$. Discuss the number of regions that the plane is divided into by the lines L_j (you may make any further assumptions that seem desirable).

Problem 4.7 A *great circle* on a sphere S in $\mathbb{R}^3$ is the intersection of S with a plane that passes through the centre of S. Suppose that a set of n great circles, no three of which meet at a point on the sphere, divides S into $X(n)$ regions. Find $X(n)$.

For more problems of this type, see [11].

5

Crossing squares

5.1 The problem

We shall work in $\mathbb{R}^2$. As usual, $[a, b] = \{x \in \mathbb{R} : a \leqslant x \leqslant b\}$, and

$$[a, b] \times [c, d] = \{(x, y) \in \mathbb{R}^2 : a \leqslant x \leqslant b, c \leqslant y \leqslant d\}.$$

When we speak of *the diagonal* of the rectangle $[a, b] \times [c, d]$, we shall always mean the diagonal from (a, c) to (b, d). Also, we shall call the square $[a, a + 1] \times [c, c + 1]$ a *unit square* when a and c are integers. Now divide the rectangle $[0, 5] \times [0, 2]$ into ten unit squares in the obvious way (as illustrated in Figure 5.1). Clearly, the diagonal of this rectangle crosses six of these unit squares.

Problem 5.1 Let p and q be positive integers and divide the rectangle $[0, p] \times [0, q]$ into pq unit squares in the obvious way. *How many of the pq unit squares are crossed by the diagonal D of the rectangle $[0, p] \times [0, q]$?*

Before we attempt to solve this, we must consider what we mean by '*crossed by the diagonal*'. Consider, for example, the square $[0, 2] \times [0, 2]$ divided into four unit squares. Does the diagonal D cross two or four unit squares? If D meets a unit square S *only* at a single vertex, does it cross S or not? Clearly, the problem is ambiguous, and this ambiguity (that is, the meaning of 'crossed') must be resolved *before* we try to solve the problem. This situation is not unusual, for mathematics demands a *much higher level* of precision than is normally used in a language. In any event, there is a decision here that *we* (the solver)

Figure 5.1 The diagonal crosses six squares

must take before moving on. In order to distinguish between the two possibilities, we shall consider a unit square S, and say that a line L

- *crosses* S if $L \cap S$ is a segment of positive length, and
- *meets* S if $L \cap S$ is non-empty (and possibly a single point).

In particular, the diagonal of $[0, 2] \times [0, 2]$ crosses exactly two of the unit squares, but meets all four of them. Now that we have clarified the ambiguity, and introduced appropriate terminology, we can move on and consider the original problem.

5.2 Lattice points on the diagonal D

A point with integer co-ordinates is called a *lattice point*, and the diagonal D of $[0, p] \times [0, q]$ contains the lattice points $(0, 0)$ and (p, q). It is important to understand when D does, or does not, contain any other lattice points.

Problem 5.2 When does the diagonal D of $[0, p] \times [0, q]$ pass through a lattice point other than its endpoints $(0, 0)$ and (p, q)?

Suppose that (u, v) is a lattice point on D; then (as D lies on the line given by the equation $yp = qx$) we have $pv = qu$, where $0 \leqslant u \leqslant p$ and $0 \leqslant v \leqslant q$. Now let $d = \gcd(p, q)$ (possibly, $d = 1$ here), and write $p = dp_1$ and $q = dq_1$. Then $p_1 v = q_1 u$, and as p_1 and q_1 are coprime, we see that p_1 divides u, and q_1 divides v. Thus $(u, v) = (p_1 t, q_1 t)$ for some integer t, and we have proved the following result.

Theorem 5.1 *The set of lattice points on the diagonal D of $[0, p] \times [0, q]$ is*

$$\{(0, 0), (p_1, q_1), (2p_1, 2q_1), \ldots, (dp_1, dq_1)\}, \tag{5.1}$$

where $d = \gcd(p, q)$, $p_1 = p/d$ and $q_1 = q/d$.

This result includes (as the special case when $d = 1$) the fact that p and q are coprime *if and only if* the only lattice points on D are its endpoints $(0, 0)$ and (p, q).

5.3 The general solution

We begin with the assumption that *p and q are coprime*; thus D does not contain any lattice points other than $(0, 0)$ and (p, q). Now consider the rectangle $[0, p] \times [0, q]$ divided into pq unit squares and label each of these squares by the sum $r + s$ of the co-ordinates (r, s) of its upper right-hand corner; thus the unit square $[m, m + 1] \times [n, n + 1]$ is labelled $m + n + 2$. As we move along the diagonal from $(0, 0)$ to (p, q), we pass from a square labelled $r + s$ to a square that has label $r + s + 1$, for we know that D does not pass through the lattice point (r, s) (and hence not into the square with label $r + s + 2$). In other words, as we pass from one square to the next, the label is increased by 1. Thus, as we move from the first square (with label 2) to the last square (with label $p + q$) we must have passed through squares with labels $2, 3, \ldots, p + q$; in other words, through $p + q - 1$ squares. Thus we have proved the following result (which the reader should confirm experimentally by drawing several pictures).

Theorem 5.2 *If* $\gcd(p, q) = 1$, *then the diagonal D of* $[0, p] \times [0, q]$ *crosses exactly $p + q - 1$ unit squares.*

Let us now consider the general case with the notation p, q, d, p_1 and q_1 as above. As $\gcd(p_1, q_1) = 1$, the diagonal D_0 of the rectangle $[0, p_1] \times [0, q_1]$ crosses $p_1 + q_1 - 1$ unit squares. By exactly the same argument, the diagonal of $[p_1, 2p_1] \times [q_1, 2q_1]$ also crosses $p_1 + q_1 - 1$ unit squares. Similarly, the diagonal of $[2p_1, 3p_1] \times [2q_1, 3q_1]$ crosses $p_1 + q_1 - 1$ unit squares, and so on. Since all of these diagonals lie on D, we see that the diagonal D of $[0, dp_1] \times [0, dq_1]$ crosses $d(p_1 + q_1 - 1)$ unit squares. Thus we have proved the following generalisation of Theorem 5.2.

Theorem 5.3 *The diagonal D of* $[0, p] \times [0, q]$ *crosses exactly $p + q - \gcd(p, q)$ of its unit squares.*

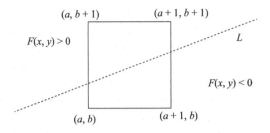

Figure 5.2 L crossing $[a, a + 1] \times [b, b + 1]$

It is easy to write a computer program to count the squares crossed by the diagonal (and so obtain numerical confirmation of our result). The diagonal D lies on the line L given by $F(x, y) = 0$, where $F(x, y) = py - qx$. The region above L is given by $F(x, y) > 0$, and the region below by $F(x, y) < 0$. Now L crosses the unit square $[a, a+1] \times [b, b+1]$ if and only if $(a, b+1)$ is above L, and $(a+1, b)$ is below L (see Figure 5.2); that is, if and only if $-p < pb - qa < q$.

We can now write a computer program to count the number of squares crossed. Roughly speaking, the program is

```
C=0
input p, input q
for a=0,1,...,p-1, for b=0,1,...,q-1
if -p < pb-qa < q then C=C+1,
```
where C counts the number of squares crossed.

Problem 5.3 Give a formula for the number of squares that the diagonal D *meets* (possibly only in a single vertex).

5.4 An alternative approach

In this section, we again assume (for simplicity) that p and q are coprime. We consider a sequence $S_1, \ldots, S_m$ of unit squares that has the property that, for each j, S_j and S_{j+1} have a common side which is either the top side, or the right-hand side, of S_j. Such a sequence represents a 'path of squares' always moving upward or to the right. We can represent this path of squares by a 'word' in the letters R and A formed as follows. If S_2 lies to the *right* of S_1, we begin our word with the letter R; if S_2 lies *above* S_1, we begin our word with A. If S_3 lies to the right of S_2, the second letter in our word is R; if it lies above S_2, the second letter in the word is A. We continue in this way, and each path of squares $S_1, \ldots, S_m$ can be represented uniquely by the initial square S_1 and a word of length $m - 1$ in the letters A and R.

Now suppose that the sequence $S_1, \ldots, S_m$ represents a path of squares that joins the bottom left-hand unit square of the rectangle $[0, p] \times [0, q]$ to its top right-hand unit square, and let W be the word which represents this path of squares. Clearly, exactly $p - 1$ of the letters in W must be R, and exactly $q - 1$ letters must be A; thus $m - 1 = (p - 1) + (q - 1)$, or $m = p + q - 1$. This agrees with our earlier result; however, it says much more than that because *it does not require the squares in the path to be placed along the diagonal.*

Let us look at this argument in another way. It is obvious that if the word $* * *AR * **$ represents a path of squares, then so does the word $* * *RA * **$, where we have changed AR to RA. Since this operation does not change the number of squares in the sequence, we can repeat this operation as often as necessary until we obtain the word $R \ldots RA \cdots A$. This word corresponds to the path of squares that runs along the bottom of the rectangle, and then up the right-hand side of the rectangle, and clearly there are $p + q - 1$ squares in this path. In fact, *any* sequence of squares that joins the bottom left square to the top right square in the prescribed way *must* contain $p + q - 1$ squares, irrespective of whether p and q are coprime or not, and irrespective of whether the squares lie on the diagonal D or not. This argument shows that *the diagonal D is perhaps not as important as we might have thought.*

5.5 A combinatorial proof

Here is a combinatorial proof of Theorem 5.3, without assuming that p and q are coprime. Imagine travelling along D from $(0, 0)$ to (p, q), and suppose that D crosses a unit square S. Then we give S the label V if D leaves S on its *vertical* (right-hand) side (including its two vertices), and the label H if D leaves S on its (top) *horizontal* side (including its two vertices). Note that we only give labels to the squares crossed by D, and that there are p squares labelled V, and q squares labelled H. Note that in this situation, the last square (and perhaps some other squares) will have *both* labels H and V. Indeed, a square S will have both labels if and only if the diagonal D exits S at its top right-hand corner, and then this must be one of the points (kp_1, kq_1), $k = 1, 2, \ldots, d$. In other words, *there are exactly d squares that have both labels.*

Let $\mathcal{V}$ and $\mathcal{H}$ be the collection of squares labelled by V and H, respectively, and let $|E|$ denote the number of elements in a set E. Then, as above, $|\mathcal{V}| = p$, $|\mathcal{H}| = q$ and $|\mathcal{V} \cap \mathcal{H}| = d$. However, the number of squares crossed by D is $|\mathcal{V} \cup \mathcal{H}|$, and it is well known that (in general) $|\mathcal{V} \cup \mathcal{H}| = |\mathcal{V}| + |\mathcal{H}| - |\mathcal{V} \cap \mathcal{H}|$. Thus $|\mathcal{V} \cup \mathcal{H}| = p + q - d$ and the proof is complete.

Problem 5.4 Consider a similar problem with parallelograms, or equilateral triangles, instead of squares.

The answer for parallelograms is the same as for squares since a shear will convert the tesselation by parallelograms into one by rectangles, and lines go to lines, intersections to intersections and so on.

5.6 The problem in three dimensions

Now we have fully understood the problem in two dimensions, we can tackle the corresponding problem in three dimensions. Note, however, that had we started with the problem in three dimensions it would probably have seemed impossibly difficult.

Problem 5.5 Let p, q and r be positive integers, and divide the cuboid $[0, p] \times [0, q] \times [0, r]$ into pqr non-overlapping unit cubes. How many of these unit cubes are crossed by the diagonal D from $(0, 0, 0)$ to (p, q, r)?

We have seen how $\gcd(p, q)$ played a crucial role in our discussion of the two-dimensional case, so the greatest common divisor must play a role in higher dimensions. However, the problem is now more complicated; for example, in the three-dimensional case we now have *four* greatest common divisors to consider, namely $\gcd(p, q)$, $\gcd(p, r)$,

$\gcd(q, r)$ and $\gcd(p, q, r)$. If, for example, $(p, q, r) = (6, 10, 15)$ then $\gcd(p, q, r) = 1$ but no two of p, q and r are coprime.

One of the earlier proofs in the case of two dimensions was based on the fact that $|A \cup B| = |A| + |B| - |A \cap B|$. If we were to try to generalise this proof in order to solve the problem in three dimensions, we would presumably have to use the fact that

$$|A \cup B \cup C| = |A| + |B| + |C| - |A \cap B| - |A \cap C| - |B \cap C| + |A \cap B \cap C|.$$

So, is it reasonable to make the following conjecture?

Conjecture 5.4 *Suppose that p, q and r are positive integers. Then the diagonal D of $[0, p] \times [0, q] \times [0, r]$ which joins $(0, 0, 0)$ to (p, q, r) crosses exactly*

$$p + q + r - \gcd\{p, q\} - \gcd\{p, r\} - \gcd\{q, r\} + \gcd\{p, q, r\}$$

of the unit cubes.

Problem 5.6 Can you write a computer program to check numerically whether they might be true? Can you decide whether this conjecture is true or not?

If this conjecture were true then it would imply the following corollary.

Corollary 5.5 *Let p, q and r be positive integers such that p, q and r are pairwise coprime. Then the diagonal D of $[0, p] \times [0, q] \times [0, r]$ which joins $(0, 0, 0)$ to (p, q, r) crosses exactly $p + q + r - 2$ of the unit cubes.*

In fact, it is not difficult to prove Corollary 5.5 directly. First, observe that (as in the two-dimensional case), if we draw a path given by $(x(t), y(t), z(t))$ that stays in the cuboid, and passes from $(0, 0, 0)$ to (p, q, r) without passing through any of the edges of the cubes (apart, of course, at the points $(0, 0, 0)$ and (p, q, r)), and is such that $x(t)$, $y(t)$ and $z(t)$ are strictly increasing functions of t, then this path must cross through exactly $p + q + r - 2$ unit cubes. Now notice that if the diagonal D passes through a vertical edge of the cube then, by projecting the entire situation vertically onto the rectangle $[0, p] \times [0, q]$, we see that we must have $\gcd(p, q) > 1$. A similar statement holds for both types of horizontal edges, so we have now proved Corollary 5.5.

This proof suggests that there is an important geometric insight to be made here. Suppose, for example, that the diagonal D does not pass through any lattice point other than its endpoints but, nevertheless, passes through (at least) one of the *edges* of the unit cubes. Then, by considering the projections onto the three co-ordinate planes, it is clear that one of the integers $\gcd(p, q)$, $\gcd(q, r)$ and $\gcd(r, p)$ *will be greater than* 1 even if $\gcd(p, q, r) = 1$.

Finally, perhaps the reader should start a serious investigation of the three-dimensional case by noting the following result.

Theorem 5.6 *The set of lattice points on the diagonal D of the cuboid $[0, p] \times [0, q] \times [0, r]$ which joins $(0, 0, 0)$ to (p, q, r) is precisely the set $\{k(p_1, q_1, r_1) : k = 0, 1, \ldots, d\}$, where $p_1 = p/d$, $q_1 = q/d$, $r_1 = r/d$ and $d = \gcd\{p, q, r\}$. In particular, if $\gcd(p, q, r) = 1$, then there are no lattice points on D other than its two endpoints.*

We illustrate the importance of using all parts of mathematics by giving a proof based on group theory, and *the reader is challenged to find an alternative proof.*

Proof It is sufficient to show that if $\gcd(p, q, r) = 1$ then the only lattice points (in $\mathbb{R}^3$) on D are its endpoints $(0, 0, 0)$ and (p, q, r). Let L be the straight line that contains D, and let $\mathcal{L}$ be the set of lattice points on L. Then $\mathcal{L}$ is an additive group (the reader should check this) that contains only isolated points (this is obviously true). Now such a group must be cyclic (again, the reader should prove this) generated, say, by (u, v, w), where u, v and w are positive integers. Thus, for some integer k, $(p, q, r) = k(u, v, w)$. However, as $\gcd(p, q, r) = 1$ we see that $k = 1$; hence $\mathcal{L}$ is generated by (p, q, r). The desired result follows immediately from this. $\qquad\square$

Problem 5.7 Can you make a conjecture about the n-dimensional problem?

6

Repeated vector products

6.1 The problem

This problem is about products of vectors lying in Euclidean space $\mathbb{R}^n$, where $n = 1, 2, 3, 4$. The problem for $\mathbb{R}^1$ (the real line) is this: take any real numbers x and a, and then repeatedly multiply x on the left by a to create the sequence $a^n x$, $n = 0, 1, 2, \ldots$.

Problem 6.1 Give a *complete* discussion of the limiting behaviour of the sequence $a^n x$, $n = 0, 1, 2, \ldots$ (that is, for *all possible values* of a and x).

Now consider the space $\mathbb{R}^2$. We can identify $\mathbb{R}^2$ with the complex plane $\mathbb{C}$ and then consider the same problem as above. Given complex numbers z and a, we form the sequence $z, az, a^2z, a^3z, \ldots$. We leave the reader to describe the limiting behaviour of this sequence in the various cases.

Problem 6.2 Suppose that $z = 1$ and $a = e^{i\pi\theta}$. Discuss *in detail* the difference between the two cases: (i) θ is rational and (ii) θ is irrational. In particular, can the sequence a^n, $n = 1, 2, \ldots$, accumulate at every point of the unit circle? When (if ever) does the sequence a^n accumulate at every point of the unit circle?

6.2 Vector products

We shall now consider vectors in $\mathbb{R}^3$. We shall assume that the reader is familiar with the vector product $\mathbf{a} \times \mathbf{b}$, the vector triple product

$\mathbf{a} \times (\mathbf{b} \times \mathbf{c})$ and the scalar triple product $\mathbf{a} \cdot (\mathbf{b} \times \mathbf{c})$ of vectors $\mathbf{a}$, $\mathbf{b}$ and $\mathbf{c}$ in $\mathbb{R}^3$. In particular, we recall that

$$\mathbf{a} \times (\mathbf{b} \times \mathbf{c}) = (\mathbf{a} \cdot \mathbf{c})\mathbf{b} - (\mathbf{a} \cdot \mathbf{b})\mathbf{c},$$

and, with the usual notation, that

$$\mathbf{a} \cdot (\mathbf{b} \times \mathbf{c}) = \begin{vmatrix} a_1 & a_2 & a_3 \\ b_1 & b_2 & b_3 \\ c_1 & c_2 & c_3 \end{vmatrix}.$$

Choose any vector $\mathbf{a}$ and, for any vector $\mathbf{x}$, consider the sequence

$$\mathbf{x}, \ \mathbf{a} \times \mathbf{x}, \ \mathbf{a} \times (\mathbf{a} \times \mathbf{x}), \ldots$$

of vectors (obtained by repeated 'left multiplication' by the vector $\mathbf{a}$). More formally, we define $\mathbf{x}_0, \mathbf{x}_1, \ldots$ inductively by

$$\mathbf{x}_0 = \mathbf{x}, \quad \mathbf{x}_{n+1} = \mathbf{a} \times \mathbf{x}_n, \quad n = 1, 2, \ldots.$$

Problem 6.3 Investigate the nature of the sequence $\mathbf{x}_n$. If you observe anything significant, write it *as a formal statement*, and then prove it.

You might choose two specific vectors $\mathbf{a}$ and $\mathbf{x}$ and start computing the sequence $\mathbf{x}_n$. As you will see, this takes a lot of time, and you may not notice anything from this mass of calculations. Perhaps we need an insight into vector products that goes beyond mere calculations? Here, we shall consider two interpretations of the vector product, one geometric and the other algebraic.

6.3 A geometric approach

A vector $\mathbf{x}$ has both *magnitude*, which we denote by $\|\mathbf{x}\|$, and *direction*. The geometric definition of the vector product $\mathbf{a} \times \mathbf{b}$ is the vector whose magnitude is $\|\mathbf{a}\| \|\mathbf{b}\| \sin\theta$, where θ is the angle between the directions of $\mathbf{a}$ and $\mathbf{b}$, and whose direction is orthogonal to the directions of $\mathbf{a}$ and $\mathbf{b}$ such that the three vectors $\mathbf{a}$, $\mathbf{b}$ and $\mathbf{a} \times \mathbf{b}$, *in this order*, form a 'right-handed' system of co-ordinates. Of course, this is *not* a complete definition until we have defined (mathematically) what we mean by the (physical) idea of a 'right-handed' system of co-ordinates. Nevertheless, we shall use this informal idea to give us some insight into the problem, and then attempt to prove the results that we find by using the more rigorous algebraic definition of a vector product.

Problem 6.4 Can you use the geometric definition of a vector product to describe the sequence x_n given above?

The geometric definition implies that x_1 is orthogonal to a, and that x_2 is orthogonal to both a and x_1. Since x_3 is orthogonal to both a and x_2, it follows that x_3 lies along the same line as x_1. Moreover, by appealing to the 'right-hand rule' we see that x_3 is in the opposite direction to x_1; thus $x_3 = \mu x_1$, where $\mu < 0$. By using the orthogonality of these vectors, we see that $\|x_3\| = \|a\|^2.\|x_1\|$, so that $x_3 = -\|a\|^2 x_1$. The reader should now be able to prove the following result.

Theorem 6.1 *Let a and x be any linearly independent vectors. Then, for $n \geqslant 1$, we have $x_{n+2} = -\|a\|^2 x_n$. In particular, for $n \geqslant 1$, the sequence of directions of x_n is periodic with period four, and if $\|a\| = 1$ then the sequence x_n itself is periodic with period four. Also, if $\|a\| < 1$ then $x_n \to 0$ as $n \to \infty$; if $\|a\| > 1$ then $\|x_n\| \to +\infty$ as $n \to \infty$.*

Having seen this result, we can now see an easier proof. First, it is usual to work with the basis $i = (1, 0, 0)$, $j = (0, 1, 0)$ and $k = (0, 0, 1)$ of mutually orthogonal vectors in $\mathbb{R}^3$, where, for example, $i \times j = k = -j \times i$. If x is a multiple of a then $x_n = 0$ for $n \geqslant 1$. Thus, from now on, we may assume that a and x are linearly independent. Then, since the vectors a, x_1 and x_2 form a right-handed triple, we may choose the co-ordinate system on $\mathbb{R}^3$ such that

$$a = \|a\|i, \quad x_1 = \|x_1\|j, \quad x_2 = \|x_2\|k,$$

where i, j and k are the usual unit vectors along the co-ordinate axes. As the reader may now check, this idea simplifies the proof.

6.4 An algebraic approach

We know that

$$a \times (x + y) = (a \times x) + (a \times y), \quad a \times \lambda x = \lambda(a \times x),$$

and these say that the map $\alpha : \mathbf{x} \mapsto \mathbf{a} \times \mathbf{x}$ is a *linear map* of $\mathbb{R}^3$ into itself. This observation (that we are dealing with linear maps) is probably an important step in the investigation *whether or not we can immediately see how to use it.* Most problems do not have a 'simple' solution, so any evidence (such as a link with linear maps) that it might have is a valuable step forward. Indeed, the recognition that this problem is about linear maps provides a compelling reason to expect it to have a simple solution.

We now know that the sequence $\mathbf{x}_n$ is obtained *by the repeated application of a linear map*, and this suggests that we should recall our knowledge of linear maps between vector spaces. This approach looks promising as a linear map can be represented by a matrix, and the composition of linear maps represented by matrices A and B is a linear map whose matrix is AB. This suggests that we can probably solve this problem by using 3×3 matrices.

It is always worthwhile to try to 'see' the problem from as many different points of view as possible, so let us consider the familiar statement that $\mathbf{a} \times \mathbf{x}$ lies in the plane orthogonal to $\mathbf{a}$. As $\alpha(\mathbf{a}) = \mathbf{a} \times \mathbf{x} = \mathbf{0}$, we see that $\mathbf{a}$ lies in the kernel of α. Thus the kernel of α is a subspace of $\mathbb{R}^3$ of dimension at least 1, and hence, by a standard theorem on vector spaces, the set of images $\alpha(\mathbf{x})$, or $\mathbf{a} \times \mathbf{x}$, taken over all $\mathbf{x}$, lies in a subspace of dimension at most 2; that is, in a plane in $\mathbb{R}^3$. It is easy to see (by algebra) that this is the plane that contains $\mathbf{0}$ and which is orthogonal to $\mathbf{a}$.

Let us use *column vectors* in $\mathbb{R}^3$. As $\mathbf{x} \mapsto \mathbf{a} \times \mathbf{x}$ is a linear map, and as linear maps are represented by matrices, for each vector $\mathbf{a}$ there is a 3×3 real matrix A such that $\mathbf{a} \times \mathbf{x} = A\mathbf{x}$. Indeed, the reader should check that

$$\mathbf{a} \times \mathbf{x} = \begin{pmatrix} 0 & -a_3 & a_2 \\ a_3 & 0 & -a_1 \\ -a_2 & a_1 & 0 \end{pmatrix} \begin{pmatrix} x_1 \\ x_2 \\ x_3 \end{pmatrix}. \tag{6.1}$$

As $\mathbf{x}_{n+1} = A^n \mathbf{x}$, we should be able to extract information about the vectors $\mathbf{x}_n$ by finding the the successive powers A^n of the matrix A. From a practical point of view, and in specific cases, the best way to find these is on a computer where we can easily compute matrix products. Finding the powers of a square matrix is an important idea in its own right, and it is often used in probability theory, and in mathematical economics, so (after revising as much of vector space theory as you need to) you should make sure that you can do this. For example, you might wish to find the eigenvalues and eigenvectors (another important idea) of A.

Problem 6.5 Choose a vector $\mathbf{a}$, compute the matrix A, and then find $A, A^2, A^3, \ldots$ by using a computer. Repeat for different values of A. Do you notice anything interesting?

Let us now return to the specific problem under consideration. The argument above suggests that we should consider the expression

$$\begin{pmatrix} 0 & -a_3 & a_2 \\ a_3 & 0 & -a_1 \\ -a_2 & a_1 & 0 \end{pmatrix}^2 \begin{pmatrix} x_1 \\ x_2 \\ x_3 \end{pmatrix}$$

and show that if $\mathbf{x}$ is orthogonal to $\mathbf{a}$, then this expression is simply $-\|a\|^2\mathbf{x}$. We leave the reader to verify this and thus to show that action of the map $\mathbf{x} \mapsto \mathbf{a} \times (\mathbf{a} \times \mathbf{x})$ on the plane orthogonal to $\mathbf{a}$ is equal to the action of the map $\mathbf{x} \mapsto -\|a\|^2\mathbf{x}$ on this plane.

Problem 6.6 You should now make sure that you can prove Theorem 6.1 (i) by the algebraic approach, (ii) by the geometric approach and (iii) by induction. It is *always* a good idea to solve a problem in as many ways as possible, for this way you will get extra insight into the problem and your solution(s).

Problem 6.7 Investigate the nature of the sequence $\mathbf{y}_n$ defined by $\mathbf{y}_1 = \mathbf{x}$ and $\mathbf{y}_{n+1} = \mathbf{y}_n \times \mathbf{a}$ (that is, multiplication by $\mathbf{a}$ on the right). How does this differ from multiplication by $\mathbf{a}$ on the left?

6.5 Quaternions

We shall now consider $\mathbb{R}^4$ which we shall identify (in the obvious way) with $\mathbb{R}^1 \times \mathbb{R}^3$. Thus, in this new setting, we redefine $\mathbf{i}$, $\mathbf{j}$ and $\mathbf{k}$ as $\mathbf{i} = (0,1,0,0)$, $\mathbf{j} = (0,0,1,0)$ and $\mathbf{k} = (0,0,0,1)$. We also define $\mathbf{1} = (1, 0, 0, 0)$, and a *quaternion* is simply a point in $\mathbb{R}^4$, say

$$x_0\mathbf{1} + x_1\mathbf{i} + x_2\mathbf{j} + x_3\mathbf{k} = (x_0, x_1, x_2, x_3),$$

where the x_j are real numbers. The *addition of quaternions* is the usual vector addition of points in $\mathbb{R}^4$, and the *multiplication of quaternions* is the natural definition derived from the requirement that

$$\mathbf{i}^2 = \mathbf{j}^2 = \mathbf{k}^2 = -1 \qquad (6.2)$$

and

$$\mathbf{ij} = \mathbf{k} = -\mathbf{ji}, \quad \mathbf{jk} = \mathbf{i} = -\mathbf{kj}, \quad \mathbf{ki} = \mathbf{j} = -\mathbf{ik}. \qquad (6.3)$$

For example, $(2\mathbf{i} + 3\mathbf{k})\mathbf{j} = 2\mathbf{k} - 3\mathbf{i}$, and $\mathbf{ijk} = -1$.

Problem 6.8 Show that (6.2) and $\mathbf{ijk} = -1$ implies (6.3).

It is clear that the set of quaternions of the form $x_0\mathbf{1} + x_1\mathbf{i}$ mirrors the algebra of the complex plane. We shall now show how the space of quaternions includes within itself the scalar and vector products in $\mathbb{R}^3$. We say that the quaternion is a *pure quaternion* if its coefficient of $\mathbf{1}$ is zero; thus it can be written in the form $(0, \mathbf{x})$, where here, $\mathbf{x} \in \mathbb{R}^3$. The reader may now show that if $\mathbf{q}_1$ and $\mathbf{q}_2$ are pure quaternions, say $\mathbf{q}_1 = (0, \mathbf{x})$ and $\mathbf{q}_2 = (0, \mathbf{y})$, then

$$\mathbf{q}_1\mathbf{q}_2 = \left(-\mathbf{x} \cdot \mathbf{y}, \mathbf{x} \times \mathbf{y} \right). \qquad (6.4)$$

In other words, *the multiplication of two pure quaternions includes both the scalar and vector products of vectors in $\mathbb{R}^3$.*

Quaternions were defined by Sir William Rowan Hamilton on 16 October 1843. Vectors were introduced *after* after quaternions, and it was equation (6.4) that prompted the American physicist J. W. Gibbs (who apparently found that 'the idea of the quaternion was quite foreign to physics') to introduce scalar and vector products in $\mathbb{R}^3$. Quaternions are now used extensively in computer graphics (particularly to describe rotations of $\mathbb{R}^3$), and in other applications, so it is perhaps time to reinstate their superiority over vectors!

7

A rolling disc

7.1 A disc rolling in a tray

Imagine a horizontal rectangular tray with vertical sides, and a disc (or coin) lying flat in the tray and rolling, *without slipping*, around the inside of the tray so that its circumference is always in contact with the edge of the tray (see Figure 7.1). We shall suppose that the disc has been rolling for all time in the past, and that it will continue to roll for all time in the future.

We may assume that the disc has radius 1, and that the sides of the tray have lengths $2 + a$ and $2 + b$, where a and b are positive (since otherwise, the disc would not fit into the tray). The motion of the disc is *periodic* if there is a point A on the circumference of the disc that coincides with a point B on a side of the tray at two (and therefore at infinitely many) different times. Given that the motion is periodic, and that A coincides with B, the *period* of the motion is the smallest number of times that the disc rolls around the tray until A next coincides with B.

Problem 7.1 Suppose that $b = \pi/2$. Show that the motion is periodic with period 1 if $a = \pi/2$, and periodic with period 2 if $a = 2\pi$. What is the period if $a = m\pi/2$ for some integer m?

Problem 7.2 Suppose that $a = 2n\pi$ for some integer n, and $b = \pi/2$. Show that the motion is periodic with period 2.

48

Figure 7.1 A disc in a tray

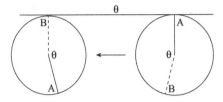

Figure 7.2 No slipping

Problem 7.3 Find conditions on a and b that imply that the motion is periodic. Given that the motion is periodic, determine the period as a function of a and b.

7.2 One circuit of the tray

In order to solve this problem we must find out how much the disc has *rotated* after it has made exactly one circuit of the tray, and to do this we must understand the phrase 'without slipping'. We may suppose that the disc moves around the boundary of the tray in an anti-clockwise direction; then the disc rotates, relative to its centre, in a clockwise direction. Consider an arc on the circumference of the disc which consists of those points on the circumference that have been in contact with one edge of the tray during some (short) time interval. Then the length of this arc is equal to the angle that the disc has rotated relative to its centre, for this is exactly what the phrase 'without slipping' means (see Figure 7.2).

Lemma 7.1 *Suppose that the disc rolls exactly once around the tray (so that the centre of the disc returns to its initial position). Then the disc has rotated clockwise by an angle $2(a + b)$ relative to its centre.*

Problem 7.4 Write down a complete proof of Lemma 7.1

7.3 Periodic motion

We now discuss periodic motion. The motion is periodic if and only
if there is some positive integer p such that, when the disc has rolled
p times around the tray (and the centre of the disc has returned to
its initial position), the disc has rotated an integral number of times,
say q times, relative to its centre. Lemma 7.1 now implies that the
motion is periodic if and only if there are integers p and q such that
$p(2a + 2b) = 2\pi q$; thus we have the following theorem.

Theorem 7.2   *The motion is periodic if and only if $a + b$ is a rational
multiple of π.*

Problem 7.5   Show that the motion is periodic if and only if $(a+b) =
k\pi/\ell$ for some coprime integers k and ℓ and that, in this case, ℓ is the
period of the motion.

Problem 7.6   Show that if a tray produces periodic motion, then so
does every tray *with the same perimeter* (but possibly with different
values of a and b).

Problem 7.7   If the motion is periodic then $a+b$ is irrational (because
π is irrational). Show, however, that the converse is false: it is possible
for $a + b$ to be irrational and the motion to be non-periodic.

There is one interesting, and important, conclusion that can be
drawn from our analysis of the rolling disc. Imagine a situation in
which the dimensions of the tray change continuously. Then, during
this process, $(a+b)/\pi$ will change infinitely often between being rational and irrational; hence, *the dynamics of the rolling disc will change
infinitely often between being periodic and non-periodic*. This is typ-
ical of many dynamical systems, and it implies that in any given *real*
situation of this type, *it is impossible to measure accurately enough to
determine whether the motion is periodic or not*!

Problem 7.8   Discuss the differences between the problem of a disc
rolling around the inside of some tray, and the same disc rolling around

the outside of the same tray. Pay particular attention to the motion of the disc at the corners of the tray.

Problem 7.9 What can you say if the disc rolls around

(1) the inside of a triangular tray?
(2) the inside of a circular tray?
(3) the outside of a triangular tray?
(4) the outside of a circular tray?

The situation in which one disc rolls around the outside of another disc is clearly important in the theory of gears. The motion of inter-linked gears is periodic, and there is an interesting history about the problem of choosing good rational approximations to an irra-tional number in order to obtain a collection of gears which yield (or approximate) a predetermined motion. This process involves contin-ued fractions, and was first examined in the context of modelling the motion of the planets.

7.4 Non-periodic motion

Is there anything interesting, or useful, that can be said about non-periodic motion? Suppose that the point A on the disc is in contact with the point B on the edge of the tray at some given time. As the motion is non-periodic, the point A will never coincide with the point B again (nor could it have coincided with B in the past). However, we can say the following.

Theorem 7.3 *Suppose that the motion is non-periodic, and that the point A on the disc is in contact with the point B on the edge of the tray at some given time. Then the point A will be within an arbitrarily small distance of B infinitely often in the future.*

Proof Let $\alpha = 2(a + b)/\pi$. As the disc makes one circuit of the tray, it will rotate by an angle $\alpha\pi$. Thus, after n circuits of the tray, the centre of the disc will have returned to its initial position, and the disc will have rotated by an angle $n\alpha\pi$ about its centre. However, as the motion is not periodic, α is irrational and now Theorem 7.3 follows immediately from the next lemma about rotations of the plane. $\quad\square$

Lemma 7.4 *Suppose that α is irrational, and let $R(z) = e^{\alpha\pi i}z$ (so that R represents the rotation of the complex plane about the origin by an angle $\alpha\pi$). Then the sequence $1, R(1), R^2(1), \ldots$ of images of the point 1 is dense on the circle $|z| = 1$.*

Proof Observe that no two of the points $1, R(1), R^2(1), \ldots$ coincide, since if they did then, for some integer p, R^p would be a rotation about the origin that fixes 1, and so would be the identity map (and then α would have to be rational). Thus these points form an infinite set on the circle $|z| = 1$, and so accumulate somewhere on the circle. It follows that there are distinct integers $p_1, p_2, \ldots$ such that $R^{p_j}(1)$ converges to some point; hence $R^{q_j}(1)$, where $q_j = p_{j+1} - p_j$, converges to 1. This means that the cyclic group G generated by R contains rotations of arbitrarily small angles (in fact, of angles $(p_{j+1} - p_j)\alpha\pi$), and since all iterates of all these rotations lie in G, it is clear that every arc of positive length on the circle $|z| = 1$ contains some point of the form $R^n(1)$. $\quad\square$

Problem 7.10 Let Y be any point on the edge of the tray that is in contact with the disc at some time (equivalently, Y is a distance at least 1 away from each corner of the tray). Show that if the motion is non-periodic, then any given point X on the circumference of the disc becomes arbitrarily close to Y infinitely often. Thus, roughly speaking, *if the motion is not periodic, then every point on the circumference of the disc comes infinitesimally close, infinitely often, to every point on the edge of the tray that is a distance at least 1 from a corner.*

The moral here is that *even in physical situations, pure mathematics is useful, perhaps even essential.* Indeed, as Lemma 7.4 has been so important here, it is worthwhile to absorb the following related results and ideas.

Theorem 7.5 *Let G be an additive subgroup of $\mathbb{R}$. Then either* (i) *G is cyclic, or* (ii) *G is dense in $\mathbb{R}$.*

Theorem 7.6 *If μ is irrational then $\{m\mu + n : m, n \in \mathbb{Z}\}$ is dense in $\mathbb{R}$.*

The proof of Theorem 7.5 As G is a subgroup of $\mathbb{R}$, we see that $0 \in G$. If $G = \{0\}$ then it is cyclic. We may now suppose that G contains a non-zero element x. As G then contains $-x$, we may assume that $x > 0$. This means that the set $\{g \in G : g > 0\}$ is non-empty, and so has a greatest lower bound a, say. If $a = 0$, then G contains arbitrary small positive elements, and then G is certainly dense in $\mathbb{R}$.

Now suppose that $a > 0$; then $G \cap (0, a) = \emptyset$. If $a \in G$ then $G = \{na : n \in \mathbb{Z}\}$, and G is cyclic with generator a. If $a \notin G$, then G contains distinct elements a_n with $a_n \to a$ but $a_n \neq a$. In this case, $a_n - a_m \in G$ for all m and n so that G contains arbitrarily small positive elements. This, however, contradicts the fact that $a > 0$ and $G \cap (0, a) = \emptyset$. □

The proof of Theorem 7.6 Let $S = \{m\mu + n : m, n \in \mathbb{Z}\}$; then S is an additive subgroup of $\mathbb{R}$. Suppose for the moment that S is cyclic, say generated by a. As 1 and μ are in S, there must be integers u and v such that $1 = ua$ and $\mu = va$, and then $\mu = v/u$ which is rational. Thus S is not cyclic and so, by Theorem 7.5, it is dense in $\mathbb{R}$. □

Problem 7.11 We know that π is irrational (this may be assumed here). By considering the additive group generated by 1 and π, show that for all α in the interval $[-1, 1]$ there exist integers $n_1, n_2, \ldots$ such that $\sin n_j \to \alpha$ as $j \to \infty$. This shows that the sequence $\sin 1, \sin 2, \sin 3, \ldots$ is dense in $[-1, 1]$.

8

Sums of powers of digits

8.1 The sum of the digits

When we write an integer n in decimal notation as $a_m \cdots a_1 a_0$, where each *digit* a_j of n is in $\{0, 1, \ldots, 9\}$, we mean that

$$n = a_0 + 10a_1 + \cdots + 10^m a_m. \tag{8.1}$$

Also, if $a_m \neq 0$ we say that n *has* $m + 1$ *digits*, or that n is an $(m + 1)$-digit number.

Problem 8.1 Show that n has m digits if and only if $10^{m-1} \leqslant m < 10^m$.

Let F be the function on the positive integers defined by $F(n) = a_0 + \cdots + a_m$; that is, $F(n)$ is the *sum of the digits of* n. For example, $F(167528) = 29$. Obviously, $F(n) = n$ if $1 \leqslant n \leqslant 9$, while $F(n) < n$ if $n > 9$. Thus if we start with a positive integer n, and then repeatedly apply the function F, we will eventually reach some integer k in $\{1, \ldots, 9\}$ and then stay at k thereafter. For brevity, we write $F^2(n) = F(F(n))$, $F^3(n) = F(F(F(n)))$ and so on; for example, if $n = 167528$ then $F(n) = 29$, $F^2(n) = 11$ and $F^3(n) = F^4(n) = \cdots = 2$.

Problem 8.2 Given n, can we predict the final value of $F^m(n)$ *without repeatedly applying* F? Also, can we estimate how many applications of F are needed to reach this value?

It should be clear that if n is very large, then the sequence $n, F(n), F^2(n), F^3(n), \ldots$ decreases rapidly. Indeed, if n has k digits then $F(n) \leqslant 9k$.

Problem 8.3 Convert this idea into an inequality of the form $F(n) \leqslant A + B \log n$ for explicit values of A and B.

Given a positive integer n, there are unique integers a and b such that $n = 9a + b$ and $1 \leqslant b \leqslant 9$ (note that we often require $0 \leqslant b \leqslant 8$ here). The value b is called the *digital root* of n, and henceforth we shall denote it by $\delta(n)$. Now $\delta(167528) = 2$, and a calculation shows that for $k \geqslant 3$, $F^k(167528) = 2$. Is this a coincidence?

Problem 8.4 Consider other values of n to obtain evidence for, or against, the suggestion that for all positive integers n, and for all sufficiently large integers m, we have $F^m(n) = \delta(n)$.

Readers who try this problem will not be surprised by the following result (which was taught to school children, long ago, in the form that if the sum $F(n)$ of the digits of a positive integer n is divisible by 9, then n itself is divisible by 9).

Theorem 8.1 *For each positive integer n, and all sufficiently large integers k, $F^k(n) = \delta(n)$.*

We shall assume that the reader is familiar with the notion of *congruence*; thus $a \equiv b \pmod{p}$ if and only if the integer p divides the difference $a - b$. Note that by definition, $\delta(n) \equiv n \pmod 9$.

The proof of Theorem 8.1 The binomial theorem implies that, for each positive integer k, $10^k = (1 + 9)^k \equiv 1 \pmod 9$. Thus if $n = a_r \cdots a_1 a_0$ then

$$n \equiv a_0 + a_1 + \cdots + a_r = F(n) \pmod 9.$$

It now follows by induction that, for all k, $F^k(n) \equiv n \pmod 9$, and, since the relation $\equiv$ is transitive, we find that for all k, $F^k(n) \equiv \delta(n) \pmod 9$. Now for sufficiently large k, both $F^k(n)$ and $\delta(n)$ are in $\{1, \ldots, 9\}$. Thus, for these k, $F^k(n) = \delta(n)$. $\qquad\square$

Problem 8.5   Show that for any positive integer n the sequence $\delta(n), \delta(n^2), \delta(n^3), \ldots$ is periodic with period at most 9. For example, when $n = 5$ the sequence is $1, 5, 7, 8, 4, 2, 1, 5, 7, 8, 4, 2, \ldots$.

8.2 The sum of the squares of the digits

In this section, we shall consider the function f defined by $f(n) = \sum_j a_j^2$; that is, $f(n)$ is the *sum of the squares of the digits of n*. We shall need to use (as above) the notation $f^k(n)$ to denote k applications of the function f; thus, for example, $f(167528) = 179$ and $f^2(167528) = f(179) = 131$.

Problem 8.6   Describe the limiting behaviour of $f^k(n)$ (as $k \to \infty$) when $n = 68$ and when $n = 89$.

Problem 8.7   Examine the sequence $n, f(n), f^2(n), \ldots$ for different values of n, and make some conjectures about the behaviour of $f^k(n)$ for all large k. Write a computer program with input n, and output the sequence $f^k(n)$ to a specified number, say N, of N terms. This exercise should provide enough evidence for a conjecture. Can you write a program that terminates once an answer is repeated?

As $f(14) = 17 > 14$, it is no longer true (as it was for the function F in the previous section) that $f(n) \leqslant n$, so this discussion is likely to be more difficult than the discussion for F. However, although it is no longer true that $n, f(n), f^2(n), \ldots$ is a decreasing sequence, it seems intuitively obvious that $f(n) < n$ *providing that n is sufficiently large*. Indeed, if n has $k + 1$ digits, then $n \geqslant 10^k$ and $f(n) \leqslant 81(k + 1)$ so that $f(n) < n$ providing that $81(k + 1) < 10^k$.

Problem 8.8   Find an integer m such that $k > m$ implies that $81(k + 1) < 10^k$.

Our next result is a refined version of this result (so the reader should attempt to prove it *before* reading the proof below).

Lemma 8.2   *If $n \geqslant 100$ then $f(n) < n$. However, $f(99) > 99$.*

Proof First, $f(99) = 162 > 99$. Next, let $n = a_k \cdots a_1 a_0$, where $k \geqslant 2$ and $a_k \geqslant 1$. Then

$$n - f(n) = a_0(1 - a_0) + a_1(10 - a_1) + \cdots + a_k(10^k - a_k).$$

Only the first term can be negative, and by checking each of the cases $a_0 = 0, 1, \ldots, 9$ we see that $a_0(1 - a_0) \geqslant -72$. Thus, if $n \geqslant 100$ then

$$\begin{aligned} n - f(n) &\geqslant a_1(10 - a_1) + \cdots + a_k(10^k - a_k) - 72 \\ &\geqslant (10^k - a_k) - 72 \\ &\geqslant (10^2 - 9) - 72 \end{aligned}$$

so that $f(n) \leqslant n - 19 < n$. $\qquad\square$

Lemma 8.2 has the obvious, but important, consequence that, for each positive integer n, there is some integer k such that $1 \leqslant f^k(n) \leqslant 99$. In fact, *there must be infinitely many k with $f^k(n)$ in $\{1, 2, \ldots, 99\}$*, since as soon as the sequence $f^k(n)$, $k = 1, 2, \ldots$, leaves this set (by taking a value greater than 99), the terms decrease again until they enter this set once again. This fact implies that, for each integer n, there are distinct integers p and q such that $f^p(n) = f^q(n)$, and this implies the next result.

Theorem 8.3 *For each integer n, the sequence $n, f(n), f^2(n), \ldots$ eventually ends in a repeating cycle.*

It only remains to check which possible cycles can arise. Clearly, it is enough to check each of the possible starting points $1, 2, \ldots, 99$ but the reader should give some thought to finding a much more efficient way to check all of these cases. With this, we have arrived at the following result.

Theorem 8.4 *For each integer n, the sequence $n, f(n), f^2(n), \ldots$ ends either with cycle $1, 1, 1, \ldots$ or with the cycle*

$$89 \to 145 \to 42 \to 20 \to 4 \to 16 \to 37 \to 58 \to 89 \cdots. \quad (8.2)$$

Problem 8.9 Given any positive integer q, study the long-term behaviour of the iterates of the map $n \mapsto a_0^q + \cdots + a_k^q$, where $a_0, \ldots, a_k$ are the digits of n. The cases $q = 1$ and $q = 2$ have been studied above.

8.3 A general result

There is a very simple, general argument which confirms Theorem 8.3
and which gives much more. Let φ be *any* function (yes, *any* function!)
from $\{0, 1, \ldots, 9\}$ to $\{1, 2, 3, \ldots\}$, and let

$$M = \max\{\varphi(0), \ldots, \varphi(9)\}.$$

Now define the function g by

$$g(n) = \sum_{j=0}^{k} \varphi(a_j), \quad n = a_k \cdots a_1 a_0.$$

As $10^m/(m+1) \to +\infty$ as $m \to \infty$, there is some integer q such that
if $m \geqslant q$ then $10^m/(m+1) > M$. Now suppose that $n \geqslant 10^q$. Then n
has $p+1$ digits, say, where $p \geqslant q$ so that $10^p > (p+1)M$. It follows
that if $n \geqslant 10^q$ then

$$g(n) = \sum_j \varphi(a_j) \leqslant (p+1)M < 10^p \leqslant n;$$

thus we have shown that if $n \geqslant 10^q$ then $g(n) < n$.

This argument shows that given *any* function φ as above, we have
$g(n) < n$ for all sufficiently large n. Thus, exactly as before, *the
sequence $n, g(n), g^2(n), \ldots$ necessarily eventually ends in a finite
repeating cycle*. In particular, we have proved the following result that
was suggested above.

Theorem 8.5 *Let m be any positive integer, and let $g(n) = a_0^m +
\cdots + a_k^m$, where $n = a_k \cdots a_1 a_0$. Then the sequence $n, g(n), g^2(n), \ldots$
eventually ends in a finite repeating cycle.*

8.4 Working in base B

The previous discussion was based on representing the positive integers in base 10. However, we can represent an integer in any integer
base B, where $B \geqslant 2$; for example, computers use binary arithmetic
(that is, base 2). Given any non-negative integer n, we write n in
base B as

$$n = a_0 + a_1 B + a_2 B^2 + \cdots + a_r B^r = (a_r \cdots a_1 a_0)_B, \qquad (8.3)$$

where these digits a_j of n satisfy $0 \leqslant a_j \leqslant B - 1$. Thus, for example,

$$253_7 = (2 \times 7^2) + (5 \times 7) + 3 = 136_{10} = 210_8 = 1021_5.$$

Problem 8.10 Discuss how much of the theory developed so far (in base 10) is valid for any base B.

In the rest of this discussion, we shall confine our attention to the function f (namely, the sum of the squares of the digits of n), except that we shall work in a general base B. As before, the key condition is to find out whether it is still true that $f(n) < n$ for all sufficient large integers n for, if so, then starting with any positive integer n, the sequence $n, f(n), f^2(n), \ldots$ eventually ends in a repeating cycle. In fact, this is true as the discussion in the previous section can easily be adapted to cover this case. Nevertheless, we shall look at this new situation in some detail. Our first result gives a sufficient condition for $f(n) > n$.

Lemma 8.6 *If n has at most three digits then $f(n)$ has at most three digits. If n has at least four digits then $f(n) < n$.*

Proof Suppose that n has at most three digits. Then $n = a_0 + a_1 B + \cdots + a_r B^r$, where $r \leqslant 2$ and $a_r \geqslant 1$. Thus

$$f(n) \leqslant (r + 1)(B - 1)^2 \leqslant 3(B - 1)^2 < [1 + (B - 1)]^3 = B^3,$$

so that $f(n)$ can have at most three digits.

To prove the second part we suppose that n has at least four digits, so that $r \geqslant 3$. We need to know

(i) that $(r + 1) \leqslant 4B^{r-3}$, and
(ii) that $4(B - 1)^2 < B^3$.

Part (i) is true because $(m + 2)/(m + 2) < 2 \leqslant B^{m+1}/B^m$, so that $B^m/(m + 1)$ is increasing with m, and (ii) is true by inspection when $B = 2$ and $B = 3$, and trivially true when $B \geqslant 4$. Thus

$$f(n) \leqslant (r + 1)(B - 1)^2 \leqslant 4B^{r-3}(B - 1)^2 < B^r,$$

so that $f(n)$ has at most r digits; hence $f(n) < n$. $\qquad\square$

Lemma 8.6 implies the following result.

Theorem 8.7 *For any n, the sequence $f^k(n)$, $k = 0, 1, 2, \ldots$, eventually end in a finite, periodic cycle of numbers, and each number in this cycle has at most three digits.*

At this point, the reader may like to experiment on a computer to obtain more data. The programs can be made more efficient as our knowledge increases, but at the moment we can at least select different values of B, and different starting points, and see what happens under repeated applications of the function f.

8.5 The fixed points of f in base B

We continue to work with the function f that is the sum of the squares of digits in base B, and we shall now examine the *fixed points* f (that is, the positive integers n such that $f(n) = n$). As 0 and 1 are the only fixed points of f when $B = 2$ (the reader should prove this), from now on we shall assume that $B \geqslant 3$.

Theorem 8.8 *Suppose that $B \geqslant 3$. Then each fixed point of f is of the form $a + bB$, and this is fixed by f if and only if*

$$(2a - 1)^2 + (2b - B)^2 = 1 + B^2. \tag{8.4}$$

The significance of Theorem 8.8 is that *it provides an algorithm for finding all fixed points of f*. Indeed, this result shows that in order to find all fixed points of f, we have only to find all possible ways of expressing $1 + B^2$ as a sum of two squares, and this very problem is discussed at great length in many texts on number theory. Perhaps the reader should now visit the library!

The proof of Theorem 8.8 Clearly, 1 is the only (positive) 1-digit fixed point of f, so we now consider a fixed point n of f that has at least two digits. By Lemma 8.6, n has at most three digits so we may write

$$a + bB + cB^2 = n = f(n) = a^2 + b^2 + c^2,$$

where $a, b, c \in \{0, 1, \ldots, B - 1\}$ and b or c is non-zero.

We shall now show that $c = 0$. First,

$$cB^2 \leqslant a + bB + cB^2 = a^2 + b^2 + c^2 \leqslant 2(B-1)^2 + c^2,$$

so that $c(B^2 - c) \leqslant 2(B-1)^2$. Now the function $h(x) = x(B^2 - x)$ is increasing on the interval $[0, B^2/2]$, which contains the interval $[0, B-1]$ so that $h(2) < h(3) < \cdots < h(B-1)$. Since $h(c) \leqslant 2(B-1)^2 < h(2)$, we see that c is 0 or 1. Now suppose, for the moment, that $c = 1$. Then $a^2 + b^2 + 1 = a + bB + B^2$. Clearly, this is false if $a = 0$, so that $a \geqslant 1$. But then $a^2 < B^2$, $b^2 < bB$ and $1 \leqslant a$ so again it is false. We deduce that $c = 0$; hence, the fixed point n satisfies $n = a+bB$, where $a^2+b^2 = a+bB$, and this is equivalent to (8.4).   □

Example 8.9 Let $B = 7$. Then it is easy to check that $1 + B^2$ can be expressed as the sum of two squares in exactly two ways, namely $1^2 + 7^2$ and $5^2 + 5^2$. Theorem 8.8 now implies that if f fixes $a + bB$, where $a, b \in \{1, \ldots, 6\}$, then

$$(2a - 1)^2 + (2b - B)^2 = 1^2 + 7^2 = 5^2 + 5^2.$$

As $2a - 1 > 0$, this means that $(2a - 1, 2b - B)$ is one of the pairs $(1, \pm 7)$, $(7, \pm 1)$ and $(5, \pm 5)$, and a little work then shows that (a, b) is one of the pairs $(1, 0)$, $(3, 1)$, $(3, 6)$, $(4, 3)$ and $(4, 4)$. Thus n, which is $a + 7b$, is one of the values 1, 10, 25, 32 and 45, and each of these is a fixed point of f. For example, $32 = (44)_7$, so that $f(32) = 4^2 + 4^2 = 32$, and $45 = (36)_7$ so that $f(45) = 3^2 + 6^2 = 45$.

Problem 8.11 Show that there are exactly three ways to write $1+82^2$ as a sum of two squares, namely $1^2 + 82^2$, $22^2 + 79^2$ and $50^2 + 65^2$. Deduce that when $B = 82$ there are exactly five fixed points f.

Problem 8.12 Show that there exactly six ways to write $1 + 132^2$ as a sum of two squares, and hence find all fixed points of f when $B = 132$.

Much more is known about this topic, see [3]. For example, it can be shown that if $1 + B^2$ is prime then $1 + B^2$ is the *only* way that it can be expressed as a sum of two squares (that is, if $1 + B^2 = a^2 + b^2$, where $a \geqslant 0$ and $b \geqslant 0$, then (a, b) is either $(1, B)$ or $(B, 1)$). This means that f has 1 as its only (positive) fixed point if and only if $1 + B^2$

is prime. As $1 + 10^2$ is prime, this explains why, when $B = 10$, f has 1 as its only fixed point. As another example, if $B = 2$ then $1 + B^2 = 5$ (a prime), so that when $B = 2$, f has 1 as its only fixed point.

To take this subject further, the reader will need to know much about those numbers that can be written as a sum of two squares, and in how many ways a given number can be written as a sum of two squares. More information on this topic can be found in [8].

9

The metric dimension

9.1 A problem in robotics

Suppose that a robot moves in a space X, and that there are a finite number of signalling devices, say $P_1, \ldots, P_n$, in X such that the robot is able to find its distance (but not its direction) from each P_j. We would like to know where to put $P_1, \ldots, P_n$ such that, by finding its distance from each P_j (and knowing the location of the P_j), the robot knows exactly where it is in X. Moreover, what is the smallest number of devices that are needed?

Let us solve this problem when X is the Euclidean plane $\mathbb{R}^2$ with the usual Euclidean distance. We shall show that *the location P of the robot is uniquely determined by its distances from any three non-collinear points U, V and W* but not, of course, by its distances from any pair of points.

Clearly, we may assume that $W = (0, 0)$; see Figure 9.1. Let $U = (u_1, u_2)$, $V = (v_1, v_2)$ and $P = (x, y)$. Then, by assumption, the distances $x^2 + y^2$, $(x - u_1)^2 + (y - u_2)^2$ and $(x - v_1)^2 + (y - v_2)^2$ are all known by the robot, as are the co-ordinates of U, V and W. It follows that

$$u_1 x + u_2 y = k_1, \quad v_1 x + v_2 y = k_2,$$

where k_1 and k_2 are known quantities. Clearly, these equations have a unique solution (x, y) (the location of the robot) providing that (u_1, u_2) and (v_1, v_2) are not scalar multiples of each other; that is, providing that U, V and W are not collinear.

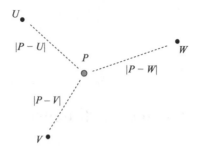

Figure 9.1 *P* determined by distances from *U*, *V* and *W*

Problem 9.1 Carry out a similar analysis when the robot moves along the real line $\mathbb{R}$, and also when it moves in $\mathbb{R}^3$. Can you solve the problem when the robot moves in $\mathbb{R}^n$?

9.2 The metric dimension of metric spaces

The problem described above makes sense for a robot moving in any set X providing that we have some notion of a distance in X, so we begin by describing exactly what we mean by a 'distance' in X. A *metric space* (X, d) is a non-empty set X with a 'distance' d which is defined for all pairs x, y of points in X, and which satisfies the following three basic requirements of a distance, namely for all x, y and z in X,

- $d(x, y) \geqslant 0$ with equality if and only if $x = y$;
- $d(x, y) = d(y, x)$;
- $d(x, z) \leqslant d(x, y) + d(y, z)$.

If we replace the Euclidean distance $|x - y|$ in the definitions of limits, continuous functions and so on, by $d(x, y)$, then these concepts make sense in the metric space. (X, d). This generalisation (from Euclidean spaces to metric spaces) is extremely powerful, and those readers who have not met this topic before should certainly spend some time learning about it now. However, we shall move on to discuss the problem of a robot moving in a metric space. In fact, these ideas are used in graph theory, chemistry, biology, robotics and many other disciplines.

Let (X, d) be a metric space, and let A be a subset of X. We say that *A resolves X* if each point x in X is *uniquely determined by the set of distances $d(x, a)$ to all of the points a in A*. Formally, this means that if $d(x, a) = d(y, a)$ for all a in A, then $x = y$; thus, in some sense, this is about a system of equations having a unique solution. Observe that if A resolves X, then we may regard the distances $d(x, a)$, where $a \in A$, as the *co-ordinates* of x with respect to the 'co-ordinate system' determined by A. Of course, the robot knows where it is in X *if and only if* A resolves X, and there are two obvious questions to ask:

(i) *what is the smallest number of points of A that are needed,*

and

(ii) *where should the points of A be located?*

The smallest number of points that resolves a metric space X is called the *metric dimension* of X and is denoted by $\beta(X)$. Of course, if $A = X$ then A certainly resolves X so that $\beta(X)$ is at most the cardinality of X. By analogy with vector spaces, if A resolves X and has cardinality $\beta(X)$ we say that A is a *metric basis* of X.

The following simple examples (which the reader should verify) should help the reader understand these ideas. For example, in (1) below, the reader is being asked to show that if $a \neq b$ then the equation $|x - a|^2 = |x - b|^2$ has a unique solution x.

Problem 9.2 (1) If $X = \mathbb{R}$ and $d(x, y) = |x - y|$, then X is resolved by any set $\{a, b\}$, where $a \neq b$, and $\beta(\mathbb{R}) = 2$.

(2) If $X = [0, +\infty)$ and $d(x, y) = |x - y|$, then X is resolved by the singleton set $\{0\}$, and $\beta(X) = 1$.

(3) If $X = (0, +\infty)$, and $d(x, y) = |x - y|$, then $\beta(X) = 2$, and a set is a metric basis of X if and only if it is of the form $\{a, b\}$, where $a \neq b$.

(4) What is $\beta(X)$ if $X = [0, 1]$ and $d(x, y) = |x - y|$?

There are many different ways to define the dimension $dim(X)$ of a set X in mathematics, and most of these have the property that $X \subset Y$ implies $dim(X) \leqslant dim(Y)$. However, the cases (2) and (3) in the previous problem show that this property is *not* always satisfied by the metric dimension.

Problem 9.3 What is $\beta(X)$ when X is the set of vertices of a regular tetrahedron? What is $\beta(X)$ when X is the set of vertices of a cube?

Problem 9.4 What is the metric dimension of the circle given by $x^2 + y^2 = 1$ in $\mathbb{R}^2$? Note that there are two questions here: (i) when we use the Euclidean distance in $\mathbb{R}^2$ and (ii) when we use the distance as the shortest path on the circle.

Problem 9.5 Let X be the subset of $\mathbb{R}^2$ comprising the union of the real axis and the circle $x^2 + (y-1)^2 = 1$, and let $d(x, y)$ be the length of the shortest arc in X that joins x to y. What is $\beta(X)$?

Problem 9.6 What is the metric dimension of the quadrant $\{(x, y) \in \mathbb{R}^2 : x > 0, y \geqslant 0\}$?

9.3 Bisectors

The idea of a bisector is useful when discussing metric dimension. Given two points, say u and v, in $\mathbb{R}^2$, the *bisector of u and v*, which we denote by $(u|v)$, is the set of points that are equidistant from u and v; that is,

$$(u|v) = \{x \in \mathbb{R}^2 : |u - x| = |v - x|\}.$$

Of course, $(u|v)$ is the Euclidean line which is usually described as an orthogonal bisector of the segment from u to v. However, the idea of a bisector is valid in any metric space (X, d), so for u and v in X, we define the *bisector* $(u|v)$ by

$$(u|v) = \{x \in X : d(x, u) = d(x, v)\}.$$

Note that if $u = v$ then $(u|v) = X$, so $(u|v)$ is only of interest when $u \neq v$. Now a subset A of X *fails to resolve* X if and only if there are distinct u and v in X such that $d(u, a) = d(v, a)$ for all a in A, and this can be expressed as follows.

Proposition 9.1 *A subset A of a metric space X resolves X if and only if it is not contained in any bisector.*

This shows that the question of finding sets that resolve (or do not resolve) X is equivalent to understanding the geometry of the bisectors in X.

Let us illustrate the use of bisectors by finding (again) the metric dimension of the Euclidean plane $\mathbb{R}^2$ with the usual Euclidean distance d. Let A be a finite subset of $\mathbb{R}^2$ and suppose that A does *not* resolve $\mathbb{R}^2$. Then (by definition) there are distinct points u and v in $\mathbb{R}^2$ such that $|u - a| = |v - a|$ for all a in A. Thus $A \subset (u|v)$, so that the A lies on a straight line. The converse, namely if A lies on a straight line then it does not resolve $\mathbb{R}^2$, is obvious; thus we have shown that A *does not resolve* $\mathbb{R}^2$ *if and only if it lies on some line* L. We state this as a theorem.

Theorem 9.2 *A subset A of $\mathbb{R}^2$ resolves $\mathbb{R}^2$ if and only if it does not lie on a line. In particular,* $\beta(\mathbb{R}^2) = 3$.

Problem 9.7 Write out a complete proof of Theorem 9.2.

Problem 9.8 Prove that a subset A of $\mathbb{R}^n$ resolves $\mathbb{R}^n$ if and only if it does not lie on some translation of some $(n - 1)$-dimensional subspace of $\mathbb{R}^n$; thus $\beta(\mathbb{R}^n) = n + 1$. Note that A resolves $\mathbb{R}^n$ if and only if a translate of A resolves $\mathbb{R}^n$ (so this is really a question about *affine* geometry).

Problem 9.9 The *open ball* $B(a, r)$ in $\mathbb{R}^3$ with centre a and radius r, where $a \in \mathbb{R}^3$ and $r > 0$, is the set $\{x \in \mathbb{R}^3 : |x - a| < r\}$. Show that $B(a, r)$ has metric dimension 4.

Problem 9.10 Let $\mathcal{C}$ be a 'solid' cube in $\mathbb{R}^3$, and let v_1, v_2 and v_3 be three vertices of $\mathcal{C}$ that lie on a single face of $\mathcal{C}$. Show that $\{v_1, v_2, v_3\}$ resolves $\mathcal{C}$, and deduce that $\beta(\mathcal{C}) \geqslant 3$.

Problem 9.11 Let X be the surface of a sphere in $\mathbb{R}^3$, and for any x and y in X, let $d(x, y)$ be the shortest distance from x to y when *measured on the surface of the sphere*. Then the bisector $(u|v)$ is the great circle on X whose plane is orthogonal to the straight line segment (in $\mathbb{R}^3$) from u to v. Deduce that a subset A of X resolves X if and only if it does not lie on a great circle, and conclude that $\beta(X) = 3$.

Problem 9.12 Discuss whether the result in the previous problem can be generalised to any dimension. Can you even *formulate* the problem carefully in $\mathbb{R}^n$? If not, why not?

9.4 Cubes and hypercubes

We shall now discuss the *edge structure* of 'cubes' or, as they are more usually called, *hypercubes*, in $\mathbb{R}^n$, where $n = 1, 2, \ldots$. We shall use the notation $\mathbf{x} = (x_1, \ldots, x_n)$, $\mathbf{y} = (y_1, \ldots, y_n)$ and so on, for points in $\mathbb{R}^n$, and $\|\mathbf{x} - \mathbf{y}\|$ and $\mathbf{x} \cdot \mathbf{y}$ for the Euclidean distance and scalar product, respectively, in $\mathbb{R}^n$. As usual,

$$\mathbf{e}_1 = (1, 0, \ldots, 0), \ \mathbf{e}_2 = (0, 1, 0, \ldots, 0), \ldots, \mathbf{e}_n = (0, \ldots, 0, 1),$$

and we shall also use the notation $\mathbf{0} = (0, \ldots, 0)$ and $\mathbf{E} = (1, \ldots, 1)$.

The *unit hypercube* in $\mathbb{R}^n$ is the set

$$Q^n = \{(x_1, \ldots, x_n) : 0 \leqslant x_j \leqslant 1, \ j = 1, \ldots, n\},$$

and the set V^n of its vertices is

$$V^n = \{(x_1, \ldots, x_n) : x_j \in \{0, 1\}, \ j = 1, \ldots, n\}.$$

Thus $(x_1, \ldots, x_n)$ is a *vertex* of Q^n if and only if each x_j is 0 or 1; equivalently, $x_j^2 = x_j$. An *edge* of Q^n is a segment in $\mathbb{R}^n$ whose endpoints are vertices $\mathbf{x}$ and $\mathbf{y}$ such that the coefficients x_j and y_j differ for *exactly one* value of j; that is, if and only if $\|\mathbf{x} - \mathbf{y}\| = 1$. For example, $(0, 0, 1, 0, 0)$ and $(0, 1, 1, 0, 0)$ are the endpoints of an edge of Q^5.

We regard the vertices and edges of Q^n as an abstract graph (that is, a collection of vertices and edges, such that two edges meet, if at all, at a common endpoint which is a vertex). The *Hamming distance* $h(\mathbf{x}, \mathbf{y})$ between two vertices $\mathbf{x}$ and $\mathbf{y}$ of Q^n is much used in graph theory (and the theory of codes), and it is the smallest number of edges that must be travelled along to get from $\mathbf{x}$ to $\mathbf{y}$; equivalently (but only for the set V^n), $h(\mathbf{x}, \mathbf{y})$ is the number of integers j such that $x_j \neq y_j$. In fact, for the set of vertices of Q^n (but not for general graphs) we have a simple formula for h, namely

$$h(\mathbf{x}, \mathbf{y}) = \sum_{j=1}^{n} |x_j - y_j| = \sum_{j=1}^{n} |x_j - y_j|^2 = \|\mathbf{x} - \mathbf{y}\|^2.$$

This formula relates the Hamming distance between two vertices to their Euclidean distance apart and hence, as we show in the next lemma, to the usual scalar product. This, in turn, will lead us to a system of linear equations which we can then solve.

Lemma 9.3 *Suppose that* **u** *and* **v** *are in* V^n. *Then*

$$h(\mathbf{u}, \mathbf{v}) = \mathbf{u} \cdot \mathbf{E} + \mathbf{v} \cdot \mathbf{E} - 2\mathbf{u} \cdot \mathbf{v}.$$

Proof Since each u_j, and v_j, is 0 or 1, we have

$$
\begin{aligned}
h(\mathbf{u}, \mathbf{v}) + 2\mathbf{u} \cdot \mathbf{v} &= \|\mathbf{u} - \mathbf{v}\|^2 + 2\mathbf{u} \cdot \mathbf{v} \\
&= \|\mathbf{u}\|^2 + \|\mathbf{v}\|^2 \\
&= \sum_j u_j^2 + \sum_j v_j^2 \\
&= \sum_j u_j + \sum_j v_j \\
&= \mathbf{u} \cdot \mathbf{E} + \mathbf{v} \cdot \mathbf{E}
\end{aligned}
$$

as required. $\square$

We shall now consider the (difficult) problem of finding the metric dimension $\beta(V^n)$ of the set V^n of vertices of the hypercube Q^n when equipped with the Hamming distance.

Problem 9.13 The hypercube Q^1 is the closed interval $[0, 1]$, $V^1 = \{0, 1\}$ and $\beta(V^1) = 1$. The hypercube Q^2 is the boundary of the unit square in $\mathbb{R}^2$, and $V^2 = \{(0, 0), (1, 0), (1, 1), (0, 1)\}$. Show that, relative to the Hamming distance, $\beta(V^2) = 2$. What is $\beta(V^3)$?

The only known values of $\beta(V^n)$ are for $n \leqslant 17$, namely

$$
\beta(V^n) = \begin{cases}
n & \text{if } n = 1, 2, 3, 4, \\
n - 1 & \text{if } n = 5, 6, 7, \\
n - 2 & \text{if } n = 8, 9, \\
n - 3 & \text{if } n = 10, 11, \\
n - 4 & \text{if } n = 12, 13, \\
n - 5 & \text{if } n = 14, 15, 16, \\
n - 6 & \text{if } n = 17,
\end{cases}
$$

and most of these have been found from a computer search. Beyond this, much work has been done on finding estimates for $\beta(V^n)$.

9.5 Linear equations

We shall now show how we can obtain information about $\beta(V^n)$ by solving systems of linear equations. We know that if $n > m$ then m linear equations in n variables have infinitely many solutions. However, in what follows we shall be considering the solutions of a linear system of equations

$$\sum_{j=1}^{n} a_{ij} z_j = 0, \quad i = 1, \ldots, m,$$

in which the coefficients a_{ij}, and the variables z_j, all lie in the set $\{-1, 0, 1\}$, which we denote by Ω. The following elementary result gives a useful, sufficient condition for a set of vertices to resolve V^n, and this enables us to obtain upper bounds on $\beta(V^n)$.

Theorem 9.4 *Let* $\mathbf{v}_1, \ldots, \mathbf{v}_m$ *be vertices of* Q^n *and suppose that the only solution* $\mathbf{z} = (z_1, \ldots, z_n)$ *of the linear system*

$$\mathbf{z} \cdot (\mathbf{E} - 2\mathbf{v}_j) = 0, \quad j = 1, \ldots, m$$

with each z_j *in* Ω, *is the trivial solution* $\mathbf{0}$. *Then* $\{\mathbf{v}_1, \ldots, \mathbf{v}_m\}$ *resolves* V^n.

Proof We suppose that the given system has only the trivial solution $\mathbf{0}$. Now take any two vertices of Q^n, say $\mathbf{x}$ and $\mathbf{y}$, and suppose that $h(\mathbf{x}, \mathbf{v}_j) = h(\mathbf{y}, \mathbf{v}_j)$ for $j = 1, \ldots, m$. Then, by Lemma 9.3,

$$(\mathbf{x} - \mathbf{y}) \cdot (\mathbf{E} - 2\mathbf{v}_j) = 0, \quad j = 1, \ldots, m. \tag{9.1}$$

Now let $\mathbf{z} = \mathbf{x} - \mathbf{y}$, and note that each z_j is in Ω. Thus, by our assumption, $\mathbf{z} = \mathbf{0}$, and hence $\mathbf{x} = \mathbf{y}$. We conclude that $\{\mathbf{v}_1, \ldots, \mathbf{v}_m\}$ resolves V^n. □

Example 9.5 Let us show that the set $\{\mathbf{v}_1, \mathbf{v}_2, \mathbf{v}_3, \mathbf{v}_4\}$, where

$$\mathbf{v}_1 = \mathbf{0}, \quad \mathbf{v}_2 = \mathbf{e}_2 + \mathbf{e}_5, \quad \mathbf{v}_3 = \mathbf{e}_3 + \mathbf{e}_5, \quad \mathbf{v}_4 = \mathbf{e}_4 + \mathbf{e}_5,$$

of vertices of Q^5 resolves V^5. By Theorem 9.4, it is sufficient to show that the equations $\mathbf{z} \cdot \mathbf{E} = \mathbf{0}$, $\mathbf{z} \cdot \mathbf{v}_2 = \mathbf{0}$, $\mathbf{z} \cdot \mathbf{v}_3 = \mathbf{0}$ and $\mathbf{z} \cdot \mathbf{v}_4 = \mathbf{0}$ or, equivalently,

$$z_1 + z_2 + z_3 + z_4 + z_5 = z_2 + z_5 = z_3 + z_5 = z_4 + z_5 = 0,$$

have $\mathbf{z} = \mathbf{0}$ as their only solution (with each z_j in Ω). Simple algebra now shows that $\mathbf{z} = (-2t, t, t, t, -t)$ for some real t and, as each z_j is in Ω, we see that $t = 0$. Thus $\mathbf{z} = \mathbf{0}$ and $\beta(V^5) \leqslant 4$.

Problem 9.14 Show that the vector space of *real* solutions $(z_1, \ldots, z_8)$ of the system

$$-z_1 + z_2 + z_3 + z_4 + z_5 + z_6 + z_7 + z_8 = 0,$$
$$z_1 - z_2 + z_3 + z_4 + z_5 + z_6 + z_7 + z_8 = 0,$$
$$z_1 + z_2 - z_3 + z_4 + z_5 + z_6 + z_7 + z_8 = 0,$$
$$z_1 + z_2 + z_3 - z_4 + z_5 + z_6 + z_7 - z_8 = 0,$$
$$z_1 + z_2 + z_3 + z_4 - z_5 + z_6 + z_7 - z_8 = 0,$$
$$z_1 + z_2 + z_3 + z_4 + z_5 - z_6 + z_7 - z_8 = 0$$

of linear equations is

$$W = \{(t, t, t, s, s, s, -2t - 2s, t - s) : t, s \in \mathbb{R}\}.$$

Deduce that the only solution of this system that lies in Ω^8 is the trivial solution $\mathbf{0}$. Now deduce that the set

$$\{\mathbf{e}_1, \mathbf{e}_2, \mathbf{e}_3, \mathbf{e}_4 + \mathbf{e}_8, \mathbf{e}_5 + \mathbf{e}_8, \mathbf{e}_6 + \mathbf{e}_8\}$$

of vertices of Q^8 resolves V^8; thus $\beta_8 \leqslant 6$.

Problem 9.15 Show that if $n \geqslant 5$, then the only solution $\mathbf{z}$ in Ω^n of the system

$$-z_1 + z_2 + \cdots + z_{n-1} + z_n = 0,$$
$$z_1 - z_2 + \cdots + z_{n-1} + z_n = 0,$$
$$\vdots$$
$$z_1 + z_2 + \cdots - z_{n-1} + z_n = 0$$

of $n - 1$ equations is the trivial solution (that is, $\mathbf{z} = 0$). Deduce that if $n \geqslant 5$ then $\{\mathbf{e}_1, \mathbf{e}_2, \ldots, \mathbf{e}_{n-1}\}$ resolves V^n; *hence if $n \geqslant 5$ then $\beta(V^n) \leqslant n - 1$.*

9.6 A lower bound for $\beta(V^n)$

It is not difficult to produce a lower bound of $\beta(V^n)$. Suppose that the set $\{\mathbf{v}_1, \ldots, \mathbf{v}_s\}$ resolves V^n. Then, by definition, the map

$$\theta : \mathbf{x} \mapsto \big(d(\mathbf{x}, \mathbf{v}_1), \ldots, d(\mathbf{x}, \mathbf{v}_s)\big)$$

is injective on V^n. As each term $d(\mathbf{x}, \mathbf{v}_j)$ is one of the numbers $0, 1, \ldots, n$, this means that θ maps V^n injectively into the product set $\{0, 1, \ldots, n\} \times \cdots \times \{0, 1, \ldots, n\}$ with s factors. As V^n has 2^n elements this shows that $2^n \leqslant (n + 1)^s$, so that

$$s \geqslant \frac{n \log 2}{\log(n + 1)}.$$

If we now assume that s takes its minimum value, we find that

$$\beta(V^n) \geqslant \frac{n \log 2}{\log(n + 1)}.$$

Problem 9.16 How does this lower bound compare with the actual values of $\beta(V^n)$ for $n = 1, \ldots, 17$?

In fact, deeper probabilistic arguments give the surprising result that

$$\beta(V^n) \sim \frac{n \log 4}{\log n} \tag{9.2}$$

as $n \to \infty$. Thus, for large n, $\beta(V^n)$ is very much smaller than n; indeed, $\beta(V^n)/n \to 0$ as $n \to \infty$.

10

Primes and irreducible elements

10.1 Primes and the integers

Let $\mathbb{Z}$ and $\mathbb{Z}^+$ be the sets of integers and positive integers, respectively. If a and b are integers, we say that a *divides* b, or that a is a *factor* of b, if and only if there is some c such that $b = ac$. If this is true, we write $a|b$. The reader is no doubt familiar with the notion of a *prime number* in $\mathbb{Z}^+$, and that primes have the following properties:

 (i) p is a prime if and only if its only factors are 1 and p;
 (ii) if p is a prime and $p|ab$ then $p|a$ or $p|b$.

Unfortunately, it is rarely mentioned until much later that (i) and (ii) have opposite characteristics: (i) is about the *factors of* a prime while (ii) is about primes *as factors*. It is hardly surprising, therefore, that in more general circumstances (i) and (ii) are not equivalent to each other. We are also familiar with the fact that in $\mathbb{Z}^+$ each integer is a product of primes, and this product is unique up to order. Since prime factorisation in $\mathbb{Z}$ is *not* unique (for example, $6 = 2 \times 3 = (-2) \times (-3)$), this also warrants further discussion. In this chapter, we shall examine these ideas in several different situations.

10.2 Primes and units in a ring

The ideas of a *prime*, an *irreducible element*, and a *unit* can be considered in any set X that has a multiplication law. The idea is to define a subset X_0 of 'special' elements of X and then try to determine whether each x in X can be written as a product of elements chosen from X_0 and, if so, to what extent this product is unique. A natural starting point for this discussion is a *commutative ring* $\mathcal{R}$; that is, a non-empty set $\mathcal{R}$ which supports the binary operations of *addition* $x + y$ and *multiplication* xy such that

(1) $\mathcal{R}$ is an abelian group with respect to addition;
(2) multiplication is associative and commutative;
(3) there is a multiplicative identity e (for all x, $xe = x = ex$);
(4) the two distributive laws hold: for all x, y and z, $(x+y)z = xz+yz$ and $z(x + y) = zx + zy$.

For example, the set of integers $\mathbb{Z}$ and the set $\mathbb{Z}_n$ of integers modulo n are rings.

We use the same terminology in $\mathcal{R}$ as we did in $\mathbb{Z}$, namely that a *divides* b, or is a *factor of* b if, for some c, $b = ac$. Next, we introduce the three most important ideas in this respect.

Definition 10.1   Let $\mathcal{R}$ be a commutative ring. Then x in $\mathcal{R}$ is

 (i) a *unit* if it has a multiplicative inverse in $\mathcal{R}$;
 (ii) a *prime* if it is not a unit, and if $x|yz$ implies $x|y$ or $x|z$;
(iii) *irreducible* if it is not 0, nor a unit, and $x = yz$ implies y or z is a unit.

Some general comments may be helpful. If u is a unit of $\mathcal{R}$, then u has a multiplicative inverse, say v, and we have $uv = vu = e$, where e is the multiplicative identity in $\mathcal{R}$. Naturally, we denote the (unique) inverse of u by u^{-1}; thus $uu^{-1} = e = u^{-1}u$. In general, if we seek a unique factorisation result, the best we can hope for is to within the insertion of units into the product; for example, if $x = yz$ and u is a unit, then we also have $x = (yu)(u^{-1}z)$. Indeed, the set $\mathbb{Z}$ of integers is a commutative ring, its units are 1 and -1 and, as we have already noted, $6 = 2{\times}3 = (-2){\times}(-3)$. More generally, we have the following important and useful fact.

Problem 10.1   Show that the set $\mathcal{U}$ of units in a commutative ring $\mathcal{R}$ is a multiplicative subgroup of $\mathcal{R}$.

The next two problems show that a prime need not be irreducible, and an irreducible element need not be a prime.

Problem 10.2 Check that $\mathbb{Z}_6$ (the set of integers modulo 6) is a commutative ring. What is the group of units in $\mathbb{Z}_6$? What are the primes in $\mathbb{Z}_6$? What are the irreducible elements of $\mathbb{Z}_6$?

Problem 10.3 Let $\mathbb{Z}[\sqrt{-5}] = \{m + n\sqrt{-5} : m, n \in \mathbb{Z}\}$. What are the units, the primes, and the irreducible elements of $\mathbb{Z}[\sqrt{-5}]$?

10.3 Gaussian integers

For a more substantial example, we consider the ring of Gaussian integers. By definition, a *Gaussian integer* is a number of the form $m + in$, where $i^2 = -1$ and $m, n \in \mathbb{Z}$. The set $\mathbb{Z}[i]$ of Gaussian integers is a ring with respect to the usual addition and multiplication of complex numbers, and the units of $\mathbb{Z}[i]$ are easily found. Indeed, if z is a unit, then there is a Gaussian integer w such that $zw = 1$. As this implies that $|z| = 1$, it is easy to see that the group of units in $\mathbb{Z}[i]$ is $\{1, -1, i, -i\}$.

Here we shall accept (without proof) the fact that in $\mathbb{Z}[i]$ (as in $\mathbb{Z}$) the concepts of prime and irreducible coincide (however, the reader is encouraged to explore this further); thus we may restrict ourselves to primes. However, to avoid confusion and to emphasise that the notion of a prime is *relative to the given ring*, we shall use the term *prime* to mean a (familiar) prime in $\mathbb{Z}$, and the term *Gaussian prime* to mean a prime in the ring $\mathbb{Z}[i]$. We know, for example, that 2 is a prime, but it is *not* a Gaussian prime because $2 = (1+i)(1-i)$, and 2 does not divide either $1 + i$ or $1 - i$. Indeed, if a and b are Gaussian integers and $a|b$, then necessarily $|a| \leqslant |b|$. However, for example, $|2| > |1 + i|$.

It is easy to produce a large number of primes in $\mathbb{Z}[i]$, for if $a + ib \in \mathbb{Z}[i]$ and $a^2 + b^2 = p$, where p is a prime, then $a + ib$ is a Gaussian prime. Indeed, suppose that $a + ib = zw$, where z and w are Gaussian integers. Then $p = a^2 + b^2 = |a + ib|^2 = |z|^2|w|^2$. Now, as p is a prime in $\mathbb{Z}$, either (i) $p = |z|^2$ and $|w| = 1$, or (ii) $p = |w|^2$ and $|z| = 1$. In case (i) we see that $w \in \{1, -1, i, -i\}$, and $(a + ib)w^{-1} = z$, so that $a + ib$ divides z. In case (ii), $a + ib$ divides w. We deduce that if $a^2 + b^2$ is prime, then $a + ib$ is a Gaussian prime. This shows, for example, that $3 + 2i$, $1 + 4i$ and $2 + 5i$ are Gaussian primes. The reader is encouraged to find more Gaussian primes and, more generally, to see how they are related to positive integers that can be written as the sum of two squares.

10.4 Irreducible matrices

Consider the collection $\mathcal{M}$ of 2×2 matrices with *non-negative integral entries* and *unit determinant*, with the usual definitions of matrix addition and multiplication. Now $\mathcal{M}$ is not a ring as discussed in Section 10.2 (why not?); nevertheless, we can still attach the same meanings to the terms *unit*, *prime* and *irreducible elements*. The 2×2 unit matrix I is the multiplicative identity of $\mathcal{M}$, and it is easy to see that this is the only unit of $\mathcal{M}$. Indeed, if A is in $\mathcal{M}$ then, since A^{-1} cannot have negative entries, $A = I$.

Let us now consider the irreducible elements of $\mathcal{M}$, so recall that A in $\mathcal{M}$ is *irreducible* if and only if it is not the zero matrix, nor the product of two matrices in $\mathcal{M}$, neither of which is the identity matrix. Now let

$$P = \begin{pmatrix} 1 & 0 \\ 1 & 1 \end{pmatrix}, \quad Q = \begin{pmatrix} 1 & 1 \\ 0 & 1 \end{pmatrix}.$$

The matrices P and Q play a special role in $\mathcal{M}$.

Theorem 10.2 *A matrix X in $\mathcal{M}$ is irreducible if and only if it is P or Q.*

Proof We shall show that Q is irreducible and *leave the reader to show that P is irreducible*. Suppose that

$$\begin{pmatrix} 1 & 1 \\ 0 & 1 \end{pmatrix} = \begin{pmatrix} a & b \\ c & d \end{pmatrix} \begin{pmatrix} p & q \\ r & s \end{pmatrix},$$

where $ad - bc = ps - qr = 1$. Then

$$\begin{pmatrix} d & d-b \\ -c & a-c \end{pmatrix} = \begin{pmatrix} d & -b \\ -c & a \end{pmatrix} \begin{pmatrix} 1 & 1 \\ 0 & 1 \end{pmatrix} = \begin{pmatrix} p & q \\ r & s \end{pmatrix}.$$

Since $0 \leqslant r = -c$, and $c \geqslant 0$ we find that $r = c = 0$. As each matrix has determinant 1, we have $ad = 1 = ps$, so that $a = d = p = s = 1$. This implies that $b + q = 1$ so that (b, q) is $(1, 0)$ or $(0, 1)$; thus Q is irreducible.

Now consider any element X of $\mathcal{M}$, say

$$X = \begin{pmatrix} a & b \\ c & d \end{pmatrix},$$

where a, b, c and d are non-negative integers with $ad - bc = 1$. First, we cannot have $b \geqslant a$ and $c \geqslant d$ else $ad = 1 + bc > bc \geqslant ad$ which is false. Next, suppose that $X \neq I$; then we cannot have $a > b$ and $d > c$ for if so, then

$$1 = ad - bc \geqslant (b = 1)(c + 1) - bc = 1 + b + c \geqslant 1,$$

so that $b = c = 0$, and $ad = 1$ so that $X = I$. We conclude that if $X \in \mathcal{M}$, and X is not the zero matrix or the identity matrix, then we must have either (i) $b \geqslant a$ and $d > c$ or (ii) $a > b$ and $c \geqslant d$. In case (i) we have, say,

$$X = \begin{pmatrix} a & b - a \\ c & d - c \end{pmatrix} \begin{pmatrix} 1 & 1 \\ 0 & 1 \end{pmatrix} = MQ.$$

Thus if X is irreducible and (i) holds, then $M = I$ so that $X = Q$. *We leave the reader to show that if X is irreducible, and (ii) holds, then $X = P$, and this completes the proof.* $\qquad\square$

Problem 10.4 Show that each X in $\mathcal{M}$ (other than I) can be written as a product of the matrices P and Q. Is this product unique? As an example, we note that

$$\begin{pmatrix} 5 & 6 \\ 4 & 5 \end{pmatrix} = \begin{pmatrix} 5 & 6 - 5 \\ 4 & 5 - 4 \end{pmatrix} Q = \begin{pmatrix} 5 & 1 \\ 4 & 1 \end{pmatrix} Q = \begin{pmatrix} 1 & 1 \\ 0 & 1 \end{pmatrix} P^4 Q = QP^4 Q.$$

We end this section by relating the ideas we have been discussing to *continued fractions* (which we assume the reader has met before); for example,

$$\frac{45}{38} = 1 + \cfrac{1}{5 + \cfrac{1}{2 + \cfrac{1}{3}}}. \tag{10.1}$$

First, we have

$$\begin{pmatrix} 45 & 13 \\ 38 & 11 \end{pmatrix} = Q^1 P^5 Q^2 P^3. \tag{10.2}$$

Now each 2×2 non-singular complex matrix represents a Möbius map; in particular, P and Q represent the maps $p(z) = z/(z+1) = 1/(1 + 1/z)$ and $q(z) = z + 1$, respectively. As $p^n(z) = 1/(n + 1/z)$ and $q^n(z) = z + n$, it follows immediately that

$$\frac{45z + 13}{38z + 11} = q^1 p^5 q^2 p^3(z) = 1 + \cfrac{1}{5 + \cfrac{1}{2 + \cfrac{1}{3 + 1/z}}}.$$

If we now put $z = \infty$, we obtain (10.1). We encourage the reader to explore these ideas in more detail.

10.5 Irreducible polynomials

Here we study the class of complex polynomials with the usual addition of polynomials, but with the *composition* (instead of the usual multiplication) of polynomials. By definition, the composition of the two polynomials u and v is $u \circ v$, where $u \circ v(z) = u\big(v(z)\big)$. This is not a commutative ring, but there is an identity I under composition, namely the polynomial $I(z) = z$, and the only polynomials that have an inverse are the linear polynomials $\ell(z) = az + b$, where $a \neq 0$.

There is a notion of irreducibility for the composition and, of course, for any polynomial p and any linear polynomial ℓ, we have $p = (p \circ \ell) \circ \ell^{-1}$, so that, in some sense, the linear polynomials play the role of units. In this context, a polynomial p is said to be *irreducible* if and only if it cannot be expressed in the form $p = u \circ v$, where u and v are both of degree at least two. Of course, if u and v are polynomials of degrees m and n, respectively, then $u \circ v$ is of degree mn, so that *any polynomial whose degree is a prime number is irreducible*.

Problem 10.5 Find an irreducible polynomial whose degree is not a prime number.

The *critical points* of a polynomial p are the (complex) solutions z of $p'(z) = 0$. The Fundamental Theorem of Algebra guarantees that if p has degree n, then there are $n-1$ critical points (counting multiple roots in the usual way), say $z_1, \ldots, z_{n-1}$. The *critical values* of p are the values $p(z_j)$, $j = 1, \ldots, n-1$, of p at these points. Informally, it seems that 'most' polynomials of degree n have $n-1$ distinct critical values, and we shall now show that all such polynomials are irreducible.

Theorem 10.3 *If a polynomial of degree n has $n-1$ distinct critical values, then it is irreducible.*

Proof We argue by contradiction, so suppose that $p(z) = u\big(v(z)\big)$, where u is a polynomial of degree r, v is a polynomial of degree s, $r \geqslant 2$, $s \geqslant 2$ and, of course, $n = rs$. Let the solutions of $p'(z) = 0$ (that is, the critical points of p) be $z_1, \ldots, z_{n-1}$, where these are listed according to their multiplicities. As $r \geqslant 2$, we see that u has a critical point, say ζ, where $u'(\zeta) = 0$. Let $w_1, \ldots, w_s$ be the solutions of $v(z) = \zeta$, again listed according to their multiplicities. Then, for each j, $p(w_j) = u\big(v(w_j)\big) = u(\zeta)$. However, we also have

$$p'(w_j) = u'\big(v(w_j)\big)v'(w_j) = u'(\zeta)v'(w_j) = 0,$$

so that every w_j is some z_i. This means that we may relabel the z_j so that $w_j = z_j$ for $j = 1, \ldots, s$. It follows that

$$\{p(z_1), \ldots, p(z_{n-1})\} = \{p(w_1), \ldots, p(w_s), p(z_{s+1}), \ldots, p(z_{n-1})\}$$
$$= \{u(\zeta), p(z_{s+1}), \ldots, p(z_{n-1})\},$$

and as this set is the set of critical values of p, and $s \geqslant 2$, we see that p has at most $n-2$ distinct critical values. We conclude that if p has $n-1$ *distinct* critical values then it is irreducible. $\square$

Problem 10.6 Let $u(z) = z^2 + a$ and $v(z) = z^2 + b$. What are the critical values of the composite polynomial $u\big(v(z)\big)$?

Problem 10.7 Show that the polynomial z^3 is irreducible but has only one critical value (so the converse to Theorem 10.3 is false).

Problem 10.8 Find an irreducible polynomial whose degree is a composite integer.

Problem 10.9 When is $z^n + a$ irreducible? When is $z^4 + az + b$, where $b \neq 0$, irreducible?

It is not difficult to show that every polynomial can be expressed as the composition of irreducible polynomials; however, *this composition may not be unique*. To see this, we define the Tchebyshev polynomials T_n as follows. For each integer n, the function $\theta \mapsto \cos(n\theta)$ is a polynomial in $\cos\theta$, and this polynomial is T_n. Thus, for all *complex numbers* θ, $\cos(n\theta) = T_n(\cos\theta)$. The lack of uniqueness in the decomposition of a polynomial into irreducible polynomials stems from the fact that

$$T_n \circ T_m = T_{m+n} = T_m \circ T_n. \tag{10.3}$$

Thus, for example, T_{16} has (at least) two irreducible decompositions, namely $T_5 \circ T_{11}$ and $T_3 \circ T_{13}$.

Problem 10.10 Prove (10.3).

11

The symmetries of a quadrilateral

11.1 The problem

What is the most symmetric quadrilateral you can think of?

Problem 11.1 Once you have answered this question (and perhaps your answer will be a 'square'), can you *create your own definition* of a 'quadrilateral' and of a 'symmetry' of a quadrilateral and then, from these definitions, *justify your answer as a rigorous piece of mathematics*?

If you cannot do this, does your answer have any real substance? Probably not. Often the main difficulty in solving a problem lies in finding, or creating, an appropriate language that enables us to discuss the problem properly. Problems can only be solved in an agreed context and, since mathematicians are free to create definitions as and when they choose, questions that are stated informally (as this one is) usually need clarification, and they may have more than one answer.

The purpose of this chapter is to emphasise *the importance of definitions in mathematics*, and that the language and context must be settled *before* we try to solve the problem. First, we must define what we mean by a *quadrilateral*, and the *symmetry* of a quadrilateral, and clearly we must address these questions *before* we attempt to answer the original question. Of course, our answer to the original problem will depend crucially on the definitions we choose.

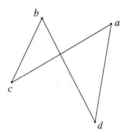

Figure 11.1 A quadrilateral with crossing sides

11.2 Quadrilaterals

So what is a quadrilateral? Is it a planar polygon with four sides, or can it be a non-planar polygon with four sides? Can it be in $\mathbb{R}^4$? Can the edges of a quadrilateral cross each other (as in Figure 11.1)? Here we shall consider a quadrilateral to be four distinct points (or vertices) a, b, c and d in either $\mathbb{R}^2$ or $\mathbb{R}^3$ (for simplicity, we ignore higher dimensions here), as well as four edges which we denote (in the obvious way) by $[a, b]$, $[b, c]$, $[c, d]$ and $[d, a]$. We shall ignore the issue of whether or not two edges cross each other since, as we shall see, this will play no part in our analysis.

11.3 Symmetries

Next, *what is a symmetry of a quadrilateral*? For example, should reflections (which cannot be achieved in the physical world) be regarded as symmetries of a quadrilateral? Notice first that the phrase 'most symmetric' implies that we must compare quadrilaterals in some way, so the next question is how are we going to do this? Here we shall say that a *symmetry* of Q is an isometry (that is, a distance-preserving map) of $\mathbb{R}^2$ (or $\mathbb{R}^3$) onto itself that maps vertices to vertices and edges to edges. With this definition, the symmetries of Q form a group (the reader should verify this), and we can use group theory to solve our problem. In fact, we can (and will) compare the symmetry groups rather than the quadrilaterals themselves. Perhaps, in some sense, the phrase 'most symmetric' means 'having the largest symmetry group', but can we always compare groups in this way? Perhaps we will obtain two groups neither of which contains the other: then what?

We denote the symmetry group of Q by $S(Q)$, and we let I denote the identity map on $\{a, b, c, d\}$. Clearly, each symmetry of Q is a permutation of V, so it can be expressed in the usual way as a product of cycles. Note, however, that as a symmetry of Q maps each side of Q to a side of Q, there are always permutations of V that are not symmetries of Q. For example, if a symmetry of Q fixes the vertex a, then it must map the set $\{b, d\}$ of vertices adjacent to a into itself, so that it must also fix the vertex c. Note also that if the four sides of Q have different lengths (and this is the generic case), then each symmetry of Q must map each side onto itself, and so map each vertex to itself. In this (generic) case, $S(Q)$ *is the trivial group* $\{I\}$.

11.4 Plane quadrilaterals

In this section, we shall assume that the quadrilateral Q lies in $\mathbb{R}^2$, where we have the usual concepts of lengths and angles available. If an isometry of $\mathbb{R}^2$ fixes three points that do not lie on a line, then it is the identity map I (the reader should prove this). Here, we shall assume that a, b, c and d do not lie on a line; then, with this assumption, *any symmetry of Q is uniquely determined by its action on the set* $\{a, b, c, d\}$ *of its vertices.*

Problem 11.2 Is an isometry of $\mathbb{R}^3$ completely determined by its action on four non-coplanar points?

It is always good to start with some examples to get the 'feel' of the problem, and it is easy to see that if Q is a square then $S(Q)$ is a group of order eight. For example, if Q has vertices $a = (1, 1)$, $b = (-1, 1)$, $c = (-1, -1)$ and $d = (1, -1)$, then $S(Q)$ consists of four rotations (including I) and the reflections across each of the four lines $x = 0$, $y = x$, $y = 0$ and $y = -x$, respectively. In terms of permutations, the eight symmetries of Q are the four rotations I, $(a\,b\,c\,d)$, $(a\,c)(d\,a)$ and $(a\,d\,c\,b)$, and the four reflections $(a\,c)(b)(d)$, $(a)(b\,d)(c)$, $(a\,d)(b\,c)$ and $(a\,b)(c\,d)$. In this case, $S(Q)$ is known as *the dihedral group D_8.*

It seems intuitively clear that, among all plane quadrilaterals, the square has the largest symmetry group so surely we should at least *conjecture that a plane quadrilateral has a symmetry group of order*

at most eight. Let us suppose, for the moment, that this is true. Our
next task must then surely be *to find all groups of order at most eight.*
This exercise is carried out in almost every basic text on group theory,
so if you don't know the answer, now is the time to visit the library.
In fact, up to isomorphisms, there is exactly one (necessarily cyclic)
group of order m, where $m = 1, 2, 3, 5, 7$; and, for non-cyclic groups,
two groups of order four, two groups of order six, and four groups of
order eight. Thus we now have *13 candidates for the symmetry group
of a given plane quadrilateral.*

Problem 11.3   Does every plane quadrilateral have a symmetry group
of order at most eight? If a quadrilateral has a symmetry group of order
eight, is it necessarily a square? Which of the 13 groups of order at
most eight arise as the symmetry group of some quadrilateral?

To complete our discussion, we should *list all possible symmetry
groups of a plane quadrilateral, and then, in each case, produce a
quadrilateral which has this as its symmetry group.* This, of course,
goes far beyond the original question, but, as always, *the original question should only be the starting point of an investigation*; equally, it
should hardly ever be the finishing point!

It is easy to prove that *the symmetry group of a plane quadrilateral Q has an order at most eight.* Suppose that Q has vertices a,
b, c and d, and sides $[a, b]$, $[b, c]$, $[c, d]$ and $[d, a]$. We say that the
vertex c is *opposite* the vertex a (because there is no side that joins
a and c), and that, for example, a and b are *adjacent* vertices. Let
σ be a symmetry of Q. As σ maps vertices to vertices, and sides to
sides, it must preserve the relations of 'opposite vertices' as well as
'adjacent vertices'. Now there are (at most) four choices for $\sigma(a)$, and
in each of these $\sigma(c)$ is uniquely determined as the vertex opposite
$\sigma(a)$. As $\sigma(b)$ is adjacent to $\sigma(a)$, there are (at most) two choices of
$\sigma(b)$, and each of these determines $\sigma(d)$. This shows that *there are at
most eight possible symmetries of Q.* In particular, this shows that the
square has the largest possible order among all symmetry groups of
quadrilaterals.

However, if we examine the permutations of $\{a, b, c, d\}$ in greater
detail, we can prove much more than this. Each permutation is a product of cycles, and *a symmetry of Q cannot, as a permutation, contain*

a three-cycle. Indeed, if it does, then it must fix the vertex, say x, not in the three-cycle; but then it must also fix the vertex opposite to x, and this cannot be so. As each group of order six contains an element of order three (check this), $S(Q)$ cannot be of order six. Since no permutation of four objects can have order five or seven, we have now proved the following result.

Theorem 11.1 *The symmetry group of a plane quadrilateral has order 1, 2, 4 or 8.*

This theorem reduces the number of potential symmetry groups (up to an isomorphism) to eight, and we must now consider which of these eight groups is the symmetry group of some quadrilateral. Could it be that there are two groups of the same order, but only one of which is the symmetry group of some quadrilateral?

Symmetry groups of order one There is only one group of order one, namely the trivial group $\{I\}$, and, as we have seen above, this is the symmetry group of a quadrilateral whose sides have different lengths.

Symmetry groups of order two
There is only one group of order two, namely the cyclic group with two elements. Now consider the following groups of order two:

$$G = \{I, (ac)\}, \quad H = \{I, (ac)(bd)\}, \quad J = \{I, (ab)(cd)\}.$$

The group G is the symmetry group of a kite whose diagonals are of different lengths; H is the symmetry group of a parallelogram that is not a rhombus or a rectangle; J is the symmetry group of a suitable trapezium with side $[a, b]$ parallel to the side $[c, d]$.

Symmetry groups of order four
There are two groups of order four, namely the cyclic group C_4 with four elements and the dihedral group D_4 with four elements. Now consider the following two groups (each isomorphic to D_4):

$$K_1 = \{I, (ac), (bd), (ac)(bd)\},$$
$$K_2 = \{I, (ad)(bc), (ab)(cd), (ac)(bd)\}.$$

Then K_1 is the symmetry group of a rhombus that is not a square, and K_2 is the symmetry group of a rectangle that is not a square.

We shall now show that C_4 *is not the symmetry group of any quadrilateral*. We argue by contradiction, so suppose that C_4 is the symmetry group of a quadrilateral Q. Then $S(Q)$ contains an element of order four. As each non-trivial symmetry of Q is either a reflection (which is of order two) or a rotation, $S(Q)$ contains a rotation, say σ, of order four. By using complex co-ordinates and choosing these appropriately, we may assume that $\sigma(z) = iz$, and that 1 is a vertex of Q. Thus the other vertices of Q are i, -1 and $-i$. However, this does *not* prove that Q is a square, for Q might have edges $[1, i]$, $[i, -i]$, $[-i, -1]$ and $[-1, 1]$. We want to show that Q has edges $[1, i]$, $[i, -1]$, $[-1, -i]$ and $[-i, 1]$. Since each edge joins two distinct points taken from $\{1, -1, i, -i\}$, there are only six possible edges, and one of these edges must be from 1 to either i or -1. We may assume that $[1, i]$ is an edge (the other case is similar). It follows that $[\sigma(1), \sigma(i)]$ is also an edge and so on, so that Q is indeed a square. In this case, $S(Q) = D_8 \neq C_4$.

Symmetry groups of order eight

The reader is invited to look at the literature on finite groups and confirm that there are exactly four groups of order eight, and that each of these contains an element of order four. With this available, the argument used above to discuss groups of order four also shows that if Q is a quadrilateral whose symmetry group is of order eight, then Q must be a square; hence $S(Q)$ is D_8.

Finally, we remark that although the reader may not be familiar with some of this group theory, it is beyond dispute that our original problem does involve group theory. It is important to realise that fact; solving problems often means that you have to learn some new mathematics!

11.5 Quadrilaterals in $\mathbb{R}^3$

Here we consider non-planar quadrilaterals in $\mathbb{R}^3$. Again, we consider Q to be given by the set V (of four vertices a, b, c and d) and E (of four edges), but now a, b, c and d are points in $\mathbb{R}^3$ which, we assume, do not lie in a plane. In this case, a *a symmetry of Q* is an isometry of $\mathbb{R}^3$ which maps $\{a, b, c, d\}$ onto itself and, of course, edges to edges.

Since there are a greater variety of isometries of $\mathbb{R}^3$ than there are of $\mathbb{R}^2$, we should expect different answers to those in the plane case.

However, some of the results proved for quadrilaterals in $\mathbb{R}^2$ remain valid for quadrilaterals in $\mathbb{R}^3$. For example, as each isometry of $\mathbb{R}^3$ is uniquely determined by its action on a set of four non-coplanar points, any symmetry of Q is uniquely determined by its action on V. This means that again each symmetry, say α, of Q can be viewed as a permutation of V, and, exactly as before, the representation of α as a product of cycles cannot contain a cycle of order three. Indeed, if α fixes one vertex, then it must also fix the opposite vertex, and so it is either the identity or a transposition. A little thought now shows that the conclusion of Theorem 11.1 is also true for quadrilaterals in $\mathbb{R}^3$. From now on we shall only concern ourselves with the case when $S(Q)$ has order eight, and we leave the other cases for the reader to examine.

Problem 11.4 Discuss the possible non-planar quadrilaterals in $\mathbb{R}^3$ (if any) that have a symmetry group of order 1, 2 or 4.

As (by assumption) $S(Q)$ has order eight, and each group of order eight contains an element of order four, we see that $S(Q)$ contains an isometry of $\mathbb{R}^3$ of order four. To make further progress, we need to find the general form of an isometry of $\mathbb{R}^3$ of order four, and this is our next task. We shall use 3×3 matrices in this discussion, and for the rest of this chapter all vectors in $\mathbb{R}^3$ will be *column* vectors. Thus if M is a 3×3 matrix and $x \in \mathbb{R}^3$, then Mx is a column vector in $\mathbb{R}^3$. We shall use the fact that, for each isometry α of $\mathbb{R}^3$, there is some a_0 in $\mathbb{R}^3$ and some 3×3 orthogonal matrix M, such that $\alpha(x) = a_0 + Mx$. If the reader is unsure of this, now is the time to study this further.

Now place a unit mass at each vertex of Q; then these masses have centre of gravity ζ, where $\zeta = (a + b + c + d)/4$. As α permutes the vertices of Q, $\{\alpha(a), \alpha(b), \alpha(c), \alpha(d)\} = \{a, b, c, d\}$; so that

$$
\begin{aligned}
\alpha(\zeta) &= a_0 + M\zeta \\
&= \tfrac{1}{4}\big[(a_0 + Ma) + (a_0 + Mb) + (a_0 + Mc) + (a_0 + Md)\big] \\
&= \tfrac{1}{4}\big(\alpha(a) + \alpha(b) + \alpha(c) + \alpha(d)\big) \\
&= \tfrac{1}{4}(a + b + c + d) \\
&= \zeta.
\end{aligned}
$$

Thus each symmetry of Q fixes ζ. Clearly, we may choose co-ordinates so that $\zeta = 0$; so now, each symmetry α of Q fixes 0. Since $a_0 = \alpha(0) = 0$, this means that $\alpha(x) = Mx$. We conclude that $S(Q)$ *is a group of order eight and of orthogonal* 3×3 *matrices.*

Now any 3×3 orthogonal matrix M has an eigenvalue 1 or -1 with corresponding eigenvector u, say. Let Π be the plane through the origin and orthogonal to u. Then M leaves Π invariant and also acts an orthogonal map of Π onto itself. By choosing co-ordinates appropriately, we may assume that u lies along the vertical axis of $\mathbb{R}^3$, so that Π is then the horizontal co-ordinate plane. Thus each symmetry of Q is now an orthogonal matrix of the form

$$M = \begin{pmatrix} p & q & 0 \\ r & s & 0 \\ 0 & 0 & \pm 1 \end{pmatrix}, \tag{11.1}$$

where the 2×2 sub-matrix

$$N = \begin{pmatrix} p & q \\ r & s \end{pmatrix} \tag{11.2}$$

is also an orthogonal matrix.

We recall (from above) that $S(Q)$ contains an element f of order four. Now the action of f is given by some matrix M as occurs in (11.1), and this means that the matrix N has order four. Thus, as we have found earlier, N is the matrix of a rotation of $\mathbb{R}^2$ order four and so by replacing f by f^{-1} if necessary, and by choosing co-ordinates appropriately, the matrix for f is one of the two matrices

$$F_1 = \begin{pmatrix} 0 & -1 & 0 \\ 1 & 0 & 0 \\ 0 & 0 & 1 \end{pmatrix}, \quad F_2 = \begin{pmatrix} 0 & -1 & 0 \\ 1 & 0 & 0 \\ 0 & 0 & -1 \end{pmatrix}. \tag{11.3}$$

If f is given by M_1, then f is a rotation of order four about the vertical co-ordinate axis. Then the vertices of Q lie in the horizontal plane (which must pass through the origin since the centre of gravity of the vertices is at the origin), and, from our discussion in Section 11.4, Q is a square lying in this plane.

If f is given by F_2 then it is a rotation R of order four about the vertical axis, followed by the reflection T across the horizontal co-ordinate plane. Such a map is called a *rotary reflection*. We now have to find all quadrilaterals Q (if any) that have a rotary reflection f of order four as a symmetry, and we may assume that f is given by the matrix F_2. The points on the vertical co-ordinate axis are fixed by f^2, so none of the vertices of Q lie on this axis. Now take any vertex of Q and choose the co-ordinates (with the same origin and vertical axis as before) such that this vertex is

$$a = \begin{pmatrix} 1 \\ 0 \\ h \end{pmatrix}.$$

Since f permutes the four vertices, and sides, of Q cyclically, we see that the four consecutive vertices of Q are

$$a = \begin{pmatrix} 1 \\ 0 \\ h \end{pmatrix}, \quad b = \begin{pmatrix} 0 \\ 1 \\ -h \end{pmatrix}, \quad c = \begin{pmatrix} -1 \\ 0 \\ h \end{pmatrix}, \quad d = \begin{pmatrix} 0 \\ -1 \\ -h \end{pmatrix}.$$

If $h = 0$ we again have a square. If $h \neq 0$, these points, and the corresponding skew quadrilateral Q, are illustrated in Figure 11.2, and it can be checked that this quadrilateral does, indeed, have a symmetry group of order eight. Perhaps we should call this quadrilateral a *skew square* since it does have four equal edges and four equal angles. Note that a, b, c and d are the four vertices of a regular tetrahedron, and the four sides of the quadrilateral are all edges of this tetrahedron.

We remark that in this discussion we have shown that any isometry of $\mathbb{R}^3$ that is of order four is either a rotation of order four about some axis or the same rotation of order four, followed by the reflection across a plane that is orthogonal to the axis of the rotation.

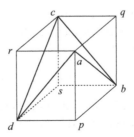

Figure 11.2 A skew square

12

Removing a vertex

12.1 Removing the vertex of a cube

Take a wooden cube, select a vertex and then, with a plane cut close to
the vertex, cut the vertex from the cube.

Problem 12.1 Given an acute-angled triangle T, can we cut the vertex from the cube so that the triangular face left exposed on the cube is
similar to T?

The reader should already be asking 'why should the question
require that T be *acute-angled*'? We might guess that this is because
any such cut will necessarily produce an acute-angled triangle, so let
us make a start on the problem by deciding whether or not this is true.
In fact, it is, and it is easy to prove. Note that although we do not know
whether this will, or will not, contribute to a solution, we should construct a proof because this can only help us improve our understanding
of the geometry in the problem.

It is easy to see whether a triangle T is acute or not in terms of
the lengths a, b and c of its sides. Indeed, by the cosine formula, T is
acute-angled if and only if

$$a^2 + b^2 > c^2, \quad b^2 + c^2 > a^2, \quad c^2 + a^2 > b^2. \tag{12.1}$$

Consider Figure 12.1 in which the three solid lines represent three
edges of the cube that meet at a vertex V, and the three dotted lines represent the edges of the triangular face $\triangle ABC$ that is exposed once V
has been cut from the cube along the plane ABC. The lengths VA, VB

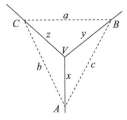

Figure 12.1 The cube with a corner cut off

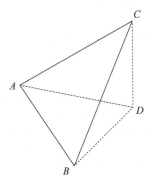

Figure 12.2 The cube with a corner cut off

and VC are denoted by the positive numbers x, y and z. Pythagoras' theorem implies that $a^2 = y^2 + z^2$, $b^2 = z^2 + x^2$ and $c^2 = x^2 + y^2$; thus (12.1) holds, and the exposed triangular face $\triangle ABC$ is indeed acute-angled.

12.2 Two geometric solutions to Problem 12.1

The first solution

Take an acute-angled triangle T made from a flat piece of wood and label the vertices A, B and C. Now place the side AB on the horizontal floor, with the vertex C (above the floor) not vertically above the side AB. Let D be the vertical projection of C on the floor and consider the tetrahedron with vertices A, B, C and D; see Figure 12.2. Because T is acute-angled, when C is almost above AB, D is almost (but not quite) on the segment AB, and the angle $\angle ADB$ is a little less than π. When the triangle is horizontal (and flat on the floor), $C = D$ and, because T is acute, $\angle ADB = \angle ACB < \pi/2$. As we rotate T about the side AB, the angle $\angle ADB$ changes continuously from being nearly π to being less than $\pi/2$. We deduce (from the Intermediate Value Theorem), that for some position of C, say C^*, we have $\angle ADB = \pi/2$. Since we always have $\angle ADC = \angle BDC = \pi/2$, we see that when $C = C^*$ we have

$$\angle ADB = \angle ADC = \angle BDC = \pi/2,$$

so that we may imagine D to be the vertex of the cube and the problem is solved.

The second solution

Given the acute-angled triangle T, label the vertices A, B and C, and construct the semicircle with diameter AB as in Figure 12.3. Now choose D on the line containing AB such that CD is orthogonal to AB. As $\angle ABC$ and $\angle BAC$ are less that $\pi/2$, C will be 'above' the segment AB, and D will be between A and B. Now let V be the point where CD meets the semicircle. As $\angle ACB < \pi/2$, and $\angle AVB = \pi/2$, V will be *inside* the triangle T. Now draw the lines AV and BV, and place T on a face of the cube so that V coincides with the vertex to be cut off, and VA and VB lie along the edges of the cube. If we now 'rotate' T about the line AB, eventually C will be vertically below V, and this solves the problem.

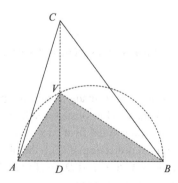

Figure 12.3 The semicircle construction

12.3 An analytic solution

Let us now briefly indicate how to solve the problem analytically. Suppose that the given triangle T has sides of lengths a, b and c. Then, as T is an acute-angled triangle, the inequalities (12.1) hold. The analysis in the previous section suggests that we should now define positive numbers x, y and z by

$$2x^2 = a^2 + c^2 - b^2, \quad 2y^2 = a^2 + b^2 - c^2, \quad 2z^2 = b^2 + c^2 - a^2.$$

Problem 12.2 Show that if x, y and z are determined as above, and if we cut the cube so that the plane of the cut passes throught the points A, B and C whose distances from the vertex V are x, y and z as in Figure 12.1, then the triangle $\triangle ABC$ is *congruent* to the triangle T.

We seemed to have proved more than we were asked to prove, so why did the original problem specify that $\triangle ABC$ need only be *similar* to T? Well, suppose that the original cube has sides of length L. If any of the numbers x, y and z are greater than L, we could not physically cut the cube in the way suggested above. On the other hand, because of the similarity condition in the problem, we can always take a sufficiently small similar 'copy' of T and then proceed as above. Of course, we can rephrase what we have proved as follows. Let $\Omega = \{(x, y, z) \in \mathbb{R}^3 : x \geqslant 0, y \geqslant 0, z \geqslant 0\}$; this is the first octant in $\mathbb{R}^3$. Then, given any acute-angled triangle T we can cut the origin $(0, 0, 0)$ from Ω with a plane cut so that the newly exposed face of Ω is *congruent* to T.

12.4 A generalisation

We can generalise the original problem as follows. Consider a tetrahedron $\mathcal{T}$ placed on a horizontal surface, and let F be the face of the tetrahedron that is on the surface. For simplicity, we shall assume that F is an equilateral triangle, and that the vertex V of $\mathcal{T}$ which is not on F lies vertically above the centroid, say Q, of F. Then the tetrahedron has rotational symmetry of order three about the vertical line VQ, and the three faces of $\mathcal{T}$ that meet at V each have the same angle θ at V. We now ask the more general question: *can we characterise, up to*

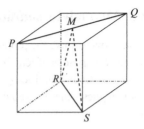

Figure 12.4 A regular tetrahedron

similarity, the set of triangular faces that are exposed after cutting V from T? This seems to be a difficult problem, even when T is a regular tetrahedron. The original problem of the cube corresponds to the case $\theta = \pi/2$, and the case of the regular tetrahedron corresponds to $\theta = \pi/3$. The tetrahedron is 'flat' when $\theta = 2\pi/3$, and, in the general case, $0 < \theta < 2\pi/3$.

One reason why the case of the cube *might be* easier than the general tetrahedron is that the *dihedral angle* of the cube (that is, the angle between two faces of the cube) is $\pi/2$. By contrast, the dihedral angle Θ between two faces of a regular tetrahedron is given by $\Theta = \cos^{-1}(1/3)$ (so that Θ is 70·528... degrees). To see this, construct a regular tetrahedron with vertices P, Q, R and S within a unit cube; see Figure 12.4. The point M is the midpoint of the segment PQ. The dihedral angle of the regular tetrahedron is the angle between the faces PQR and PQS, and this is the angle Θ, say, at M in the triangle RSM.

Problem 12.3 Show that $\cos \Theta = 1/3$.

We now ask the reader to consider the dihedral angle of a tetrahedron where all three angles at a vertex V are equal to θ.

Problem 12.4 Let T be a tetrahedron with vertex V such that all three angles at V (in the faces that meet at V) are equal to θ. Show that if Θ is the dihedral angle between any two faces meeting at V, then $2 \cos \frac{1}{2}\theta \sin \frac{1}{2}\Theta = 1$. You should first check that this formula agrees with the known results when V is the vertex of a cube, and a vertex of a regular tetrahedron.

12.5 More general tetrahedra

Consider the tetrahedron $\mathcal{T}$ described in Section 12.4. It is convenient to let $\lambda = 2\cos\theta$; then, as $0 < \theta < 2\pi/3$, we see that $-1 < \lambda < 2$. Note that $\theta < \pi/2$ if and only if $\lambda > 0$, and that $\theta > \pi/2$ if and only if $\lambda < 0$. The cube, the regular tetrahedron, and the 'flat' tetrahedron correspond to the cases $\lambda = 0$, $\lambda = 1$, and $\lambda = -1$, respectively. We can now generalise the result that, for a cube, each triangular section T is an acute-angled triangle.

Theorem 12.1 *Let $\mathcal{T}$ be a tetrahedron with vertex V, and suppose that each of the three faces of $\mathcal{T}$ that meet at V have an angle θ at V. Let T be the triangular face that is exposed when V is cut from $\mathcal{T}$. If $0 < \theta \leqslant \pi/2$ then each angle in T is at most $\pi - \theta$. If $\pi/2 < \theta \leqslant 2\pi/3$ then each angle in T is at most θ. In other words, in both cases, each angle in T is at most $\max\{\theta, \pi - \theta\}$.*

Proof We use Figure 12.1 again, but now we regard this as representing the vertex of the given tetrahedron where each of the faces has an angle θ at V. First, suppose that $0 < \theta < \pi/2$. In this case, we obtain the three equations $a^2 = y^2 + z^2 - \lambda yz$, $b^2 = z^2 + x^2 - \lambda zx$ and $c^2 = x^2 + y^2 - \lambda yx$. Now $\angle A < \pi - \theta$ if and only if $\cos\angle A > \cos(\pi - \theta) = -\lambda/2$, and this is so (from the cosine formula) if and only if $b^2 + c^2 - a^2 + \lambda bc > 0$. Now, from the equations above,

$$b^2 + c^2 - a^2 + \lambda bc = \lambda(y-x)(z-x) + (2-\lambda)x^2 + \lambda bc$$
$$\geqslant \lambda\big[bc + (y-x)(z-x)\big]$$
$$> 0,$$

since, from the triangle inequality, $|x - z| < b$ and $|x - y| < c$. This shows that $\angle A < \pi - \theta$ and, by symmetry, $\angle B$ and $\angle C$ are also at most $\pi - \theta$. We leave the reader to complete the proof in the remaining case, namely when $\theta > \pi/2$. $\qquad\square$

Finally, we remark (and the reader should prove) that if $\theta \neq \pi/2$ then the triangular section T *can have an obtuse angle*; thus the cube is *the only case* for which T is guaranteed to be an acute-angled triangle.

13

Squares within squares

13.1 Squares on a geoboard

A point (x, y) in the plane $\mathbb{R}^2$ is a *lattice point* if x and y are integers. We shall be asking questions about the geometry of polygons all of whose vertices are lattice points. A *geoboard* is a flat piece of wood with nails located at the lattice points, and polygons of the type we are interested in can easily be formed by using an elastic band around the nails.

Take a positive integer n, and consider the 'square' set

$$\mathcal{L}_n = \{(a, b) : a, b = 0, 1, 2, \ldots, n\}, \tag{13.1}$$

of $(n + 1)^2$ lattice points.

Problem 13.1 How many squares, with sides parallel to the co-ordinate axes, can be found by choosing four points from $\mathcal{L}_n$?

Problem 13.2 How many squares with vertices in $\mathcal{L}_n$ can be found if we no longer insist that the squares have their sides parallel to the co-ordinate axes?

13.2 The solutions

First, we consider Problem 13.1. Take such a square, say S, of side k, and let the location of the bottom left-hand corner of S be at (p, q) . The top right-hand corner of S is at $(p+k, q+k)$; thus we can construct such a square S if and only if

$$0 \leqslant p < p + k \leqslant n, \quad 0 \leqslant q < q + k \leqslant n,$$

or, equivalently, if p and q lie in the set $\{0, 1, \ldots, n - k\}$. Thus there are exactly $(n - k + 1)^2$ such squares of side length k, and the required answer is

$$\sum_{k=1}^{n} (n - k + 1)^2.$$

The reader should now put $m = n - 1 + k$ and then show that this sum is $\frac{1}{6} n(n+1)(2n+1)$. Now check this experimentally when, say, $n = 3$.

Problem 13.3 Now achieve the same result by creating, and then solving, a recurrence relation (consider both $\mathcal{L}_n$ and $\mathcal{L}_{n+1}$).

Now consider Problem 13.2. First, *how many squares can be formed whose vertices lie on the edge of an $n \times n$ square?* In this case, it is easy to see that there are exactly n such squares; indeed, given the lattice points in (13.1), the vertices will be at the points $(p, 0)$, (n, p), $(n - p, n)$ and $(0, n - p)$, where $p = 0, 1, \ldots, n - 1$.

We can now answer the second question. Given any such square Q there is a unique smallest square, say Q^*, which has sides parallel to the co-ordinate axes, and which is such that the vertices of Q lie on the edges of Q^*. There are $(n - k + 1)^2$ squares Q^* of side length k, where $k = 1, \ldots, n$, and each such square Q^* has exactly k squares Q whose vertices lie on the edges of Q^*. Thus the total number of squares that can be made by choosing four vertices from the set (13.1) is

$$\sum_{k=1}^{n} k(n - k + 1)^2. \tag{13.2}$$

The reader should now show that this is

$$\frac{1}{12} n(n + 1)^2 (n + 2),$$

and again check it experimentally when $n = 3$.

The results just proved suggest (indeed, imply) that, for $n = 1, 2, \ldots$, the integer 6 divides $n(n + 1)(2n + 1)$, and 12 divides $n(n + 1)^2 (n + 2)$. Can you prove this directly?

13.3 Areas of squares on a geoboard

Problem 13.4   Is it possible to construct a square on a geoboard with an area of, for example, 3,217? More generally, what are the possible areas of squares formed on a geoboard?

Suppose that a square on a geoboard has a side of length L and adjacent vertices (a, b) and (c, d). Then $L^2 = (a - c)^2 + (b - d)^2$, so the set of possible areas of squares on a geoboard is a subset of the set of positive integers that can be written as a sum of two squares. To make further progress we turn to the following theorem (which occurs in many texts on number theory).

Theorem 13.1   *A positive integer n can be written as the sum of two squares if and only if in its decomposition into a product of its prime factors, each prime of the form $4m + 3$ occurs to an even power.*

For example, $3 \times 5^4 \times 2^7$ cannot be written as a sum of two squares, whereas $3^2 \times 5^4 \times 2^7$ can. Now 3,217 is a prime, and as it is of the form $4m + 1$, it is a sum of two squares, namely $9^2 + 56^2$. Thus there is a square with area 3,217 on a (sufficiently large) geoboard. By contrast, there is no such square with area, say 51.

Problem 13.5   Find all *areas* of squares whose vertices lie in $\mathcal{L}_6$. Now find all *squares* whose vertices lie in $\mathcal{L}_6$. Which area is the area of most squares?

13.4 Areas of polygons on a geoboard

Problem 13.6   What can we say about the areas of polygons formed on a geoboard? Do they all have rational areas?

As a warning before we start this problem, we remark that a triangle with vertices at lattice points in Euclidean space $\mathbb{R}^3$ *need not have rational area.* For example, the triangle with vertices $(1, 0, 0)$, $(0, 1, 0)$ and $(0, 0, 1)$ has area $\sqrt{3}/2$. Why, then, should a triangle in the plane, with vertices at lattice points, have a rational area? Moreover, if all plane triangles with lattice points as vertices do have rational areas, then surely we must show (and certainly understand) where our argument used to prove this breaks down in $\mathbb{R}^3$. Perhaps this problem is not as simple as it may appear at first sight? Or perhaps we are overlooking something simple?

The different outcome for triangles lying in the plane compared with those that lie in space can be explained in terms of vector products. Suppose that T is a triangle in $\mathbb{R}^3$ whose vertices are lattice points in $\mathbb{R}^3$. By applying a translation (which preserves area), we may assume that the vertices of T are at the lattice points $\mathbf{0} = (0, 0, 0)$, $\mathbf{a} = (a_1, a_2, a_3)$ and $\mathbf{b} = (b_1, b_2, b_3)$, where a_i and b_j are integers. Now

$$\begin{aligned} \text{area}(T) &= \tfrac{1}{2}|\mathbf{a} \times \mathbf{b}| \\ &= \tfrac{1}{2}\sqrt{(a_2 b_3 - a_3 b_2)^2 + (a_3 b_1 - a_1 b_3)^2 + (a_1 b_2 - a_2 b_1)^2} \end{aligned}$$

and, in general, this will be irrational. However, if T lies in $\mathbb{R}^2$, then $a_3 = b_3 = 0$, and so the area of T is $\tfrac{1}{2}|a_1 b_2 - a_2 b_1|$ which is rational.

We now prove a general result about cutting (not necessarily convex) polygons into triangles, for this will enable us to focus on triangles rather than the more complicated polygons.

Theorem 13.2 *Any polygon can be decomposed into a finite number of triangles.*

Proof A polygon P has a finite number of sides, each of which determines a Euclidean line. Let these lines be $L_1, \ldots, L_m$, and consider the complement Ω of $L_1 \cup \cdots \cup L_m$. Each component of Ω is a convex polygonal region (for it lies on one side of each L_i, and so is the intersection of a finite number of open half-planes). Some of these polygonal regions will be bounded, and some will be unbounded; let the bounded regions be $R_1, \ldots, R_k$. Since the original polygon P is bounded, the interior of P must be the union of some of the R_j together with some line segments. As each R_j is a convex polygon, it can easily be decomposed into triangles (for we can select one of its vertices, and then join this vertex to each of its other vertices by straight line segments). We conclude that P can be decomposed into a finite number of triangles.   □

Theorem 13.3 *The area of a polygon formed on a geoboard is a rational number.*

Proof Consider a polygon P formed on a geoboard. Each of the lines L_j in the proof of Theorem 13.2 passes through two vertices of P, and as these are lattice points, L_j is given by an equation of the form $ax + by = c$, where a, b and c are rational. Any two such lines intersect in a rational point (that is, a point with rational co-ordinates); thus each vertex of each R_j (again see the proof of Theorem 13.2) is a rational point. This shows that P can be decomposed into a finite number of triangles, each of which has rational vertices. We have now reduced the problem to showing *that a plane triangle with rational vertices has rational area.*

Suppose then that T is a plane triangle with rational vertices. Then we can apply the map $F : (x, y) \mapsto (qx, qy)$, where q is a positive integer that is chosen so that each vertex of T maps to a lattice point. Then (see above) the image triangle has rational area, and since F changes area by a factor q^2, the original triangle T also has rational area.   □

Problem 13.7 Which rational numbers arise as the area of a triangle whose vertices are lattice points?

For more information on this topic, see [1].

13.5 Pick's Theorem

The mention of a geoboard often prompts a reference to Pick's Theorem which, as we shall now show, is sufficient to solve our problem. In 1899 the Austrian mathematician George Alexander Pick (1859–1942) proved the following result.

Theorem 13.4 (Pick's Theorem) *Let Γ be the boundary curve of a (not necessarily convex) plane polygon all of whose vertices are lattice points, and let P be the polygonal region that is inside Γ. Then the area of P is $I + \frac{1}{2}B - 1$, where I is the number of lattice points in P, and B is the number of lattice points on Γ.*

It is immediate that a polygonal region whose vertices are lattice points has rational area; indeed, the area is $N/2$, where N is the integer $2I + B - 2$. More generally, by scaling (as described above), any polygonal region with rational vertices has rational area.

Let us give a proof of Pick's Theorem. Here we shall follow the ideas used to develop measure theory: that is, we start with some basic figures, and then build the theory up to accommodate a union of such figures.

Let R be a rectangle whose vertices are lattice points, and whose sides are parallel to the co-ordinate axes, say

$$R = \{(x, y) : a \leqslant x \leqslant a + p, b \leqslant y \leqslant b + q\},$$

and let

- $A(R)$ be the area of R;
- $I(R)$ be the number of lattice points in the interior of R;
- $B(R)$ be the number of lattice points on the boundary of R.

A simple calculation shows that

$$A(R) = pq, \quad I(R) = (p - 1)(q - 1), \quad B(R) = 2p + 2q,$$

so that, indeed, we have the formula

$$A(R) = I(R) + \tfrac{1}{2}B(R) - 1. \tag{13.3}$$

The important observation here is that this formula *is independent of p and q*; thus, it holds for all such rectangles. We shall use the same functions $A(P)$, $I(P)$ and $B(P)$ for any polygonal region P whose vertices are lattice points.

Is the formula true for triangles? Consider the triangle T with vertices $(0, 0)$, $(p, 0)$ and (p, q). Then $A(T) = \tfrac{1}{2}pq$. The number $B(T)$ of lattice points on the boundary of T is $p + q + d$, where $d = \gcd\{p, q\}$. Since we can regard T as one-half of the rectangle with vertices $(0, 0)$, $(p, 0)$, (p, q) and $(0, q)$, and since we can rotate this rectangle onto itself by a rotation of angle π about the mid-point of its diagonal, we see that the number of lattice points strictly inside the rectangle is $d - 1 + 2I(T)$. As this number is also $(p - 1)(q - 1)$, we see that $d - 1 + 2I(T) = (p - 1)(q - 1)$, so that

$$2[I(T) + \tfrac{1}{2}B(T) - 1] = pq,$$

so that again (13.3) holds. Once again, the result is independent of p and q.

Pick's Theorem now follows from this and the next result.

Theorem 13.5   *Let P_1 and P_2 be non-overlapping polygonal regions, all of whose vertices are lattice points, and suppose that P_1 and P_2 have a common edge (bounded by lattice points). If $A(P_j) = I(P_j) + \frac{1}{2}B(P_j) - 1$ for $j = 1, 2$, then*

$$A(P_1 \cup P_2) = I(P_1 \cup P_2) + \tfrac{1}{2}B(P_1 \cup P_2) - 1. \qquad (13.4)$$

Proof   We suppose that the common edge is E, and that the lattice points on E are, in this order along E, $v_0, v_1, \ldots, v_{q+1}$. Now $v_1, \ldots, v_q$ each contribute $\frac{1}{2} + \frac{1}{2}$ to $A(P_1) + A(P_2)$, and each contributes the same, namely $+1$, to $A(P_1 \cup P_2)$. The vertices v_0 and v_{q+1} each contribute $\frac{1}{2} + \frac{1}{2}$ to $A(P_1) + A(P_2)$, and $\frac{1}{2}$ to $A(P_1 \cup P_2)$. Thus the pair $\{v_0, v_q\}$ contributes 2 to $A(P_1) + A(P_2)$ and 1 to $A(P_1 \cup P_2)$. However, we also have a term $-1 - 1$ in the expression for $A(P_1) + A(P_2)$, and a term -1 in the expression for $A(P_1 \cup P_2)$; thus (13.4) holds. This completes the proof of Pick's Theorem. $\square$

13.6 Higher dimensions

The problem becomes more interesting when we consider it in higher dimensions, so let us consider polyhedra whose vertices are at lattice points in $\mathbb{R}^3$.

Problem 13.8   Do all polyhedra formed on a three-dimensional geoboard have rational volumes?

Essentially, the same disection argument as that given for polygons holds for polyhedra in $\mathbb{R}^3$, so the real problem is to decide whether or not a tetrahedron in $\mathbb{R}^3$ with rational vertices has a rational volume. By a suitable translation, we may assume that the tetrahedron has vertices $\mathbf{0}, \mathbf{a}, \mathbf{b}$ and $\mathbf{c}$, each of which is a rational point in $\mathbb{R}^3$. The volume of the tetrahedron is one-sixth of the scalar triple product $|\mathbf{a}\cdot(\mathbf{b} \times \mathbf{c}|$ and, as this expression is a 3×3 determinant in the co-ordinates of $\mathbf{a}, \mathbf{b}$ and $\mathbf{c}$, the volume is rational. Thus *a polyhedron in* $\mathbb{R}^3$ *with rational vertices has rational volume.*

Problem 13.9 Choose four lattice points in $\mathbb{R}^3$ and compute the volume of the tetrahedron that has these points as vertices.

13.7 Pick's Theorem in higher dimensions

A natural question now arises: Does Pick's Theorem extend to higher dimensions? If it does, then it will surely give another solution to the problem in higher dimensions. From the discussion above, we see that the formula $I + \frac{1}{2}B - 1$ may not give the area of a triangle in $\mathbb{R}^3$. Perhaps Pick's Theorem generalises to $\mathbb{R}^3$ if we seek the volume of a tetrahedron in $\mathbb{R}^3$ instead of the area of a triangle? Even this fails. Consider, for example, a tetrahedron T whose four vertices in $\mathbb{R}^3$ are $(0, 0, 0)$, $(1, 0, 0)$, $(0, 1, 0)$ and $(1, 1, k)$, where k is a positive integer. As T lies vertically above the unit square in $\mathbb{R}^2$, the only lattice points that can lie inside or on T are the four vertices of T (regardless of the value of k). As the volume of T changes continuously with k, there can be no immediate generalisation of Pick's Theorem to the volume of an integral polyhedron in $\mathbb{R}^3$.

In fact, Pick's Theorem does generalise to higher dimensions (this was first generalised to three dimensions in [12]), and as is so often the case in mathematics, *it has to be recast in a different form before it can be generalised.* This is an important message to learn: *simple descriptions are usually too simple to describe complicated situations*!

Roughly speaking, this is the idea. Consider a polygon P (including its boundary curve) and, for a positive integer k, the lattice $\Lambda_k = \{(x/k, y/k) : x, y \in \mathbb{Z}\}$. By considering the double integral of 1 over P (which, of course, is the area of P) it is more or less self-evident that if $N(k)$ is the number of points of Λ_k that lie in P, then $N(k)/k^2 \to \text{area}(P)$ as $k \to \infty$. This shows that $N(k) \leqslant Mk^2$ for some constant M. For more information, including the n-dimensional case, see [6].

14

Catalan numbers

14.1 Introduction

The Catalan numbers $C_0, C_1, C_2, \ldots$ form a sequence of positive integers that are remarkable because they occur naturally in so many situations in which we are required to count a number of possibilities. These numbers were discovered in the 1730s in China by the Mongolian mathematician Antu Ming; they were discussed (in a different context) by the prolific Swiss mathematician Leonhard Euler around 1751 and introduced in articles in 1838 and 1839 by the Belgian mathematician Eugène Catalan. Only later they were given the name *Catalan numbers* and since then many mathematicians have contributed to our understanding of them. They are the central theme in this chapter, and for more details, we refer the reader to, for example, [7].

There are many ways to introduce the Catalan numbers. The simplest way (but without any motivation) is

$$C_0 = 1, \quad C_n = \frac{1}{n+1}\binom{2n}{n} = \binom{2n}{n} - \binom{2n}{n+1}, \quad n \geqslant 1, \quad (14.1)$$

or, equivalently (and the reader should prove the equivalence)

$$C_0 = 1, \quad C_{n+1} = \frac{2(2n+1)}{n+2}C_n. \quad (14.2)$$

However, these simple formulae conceal the real reason why the Catalan numbers are so ubiquitous. We shall see later that they satisfy the recurrence relation

$$C_n = C_0 C_{n-1} + C_1 C_{n-2} + \cdots + C_{n-2}C_1 + C_{n-1}C_0, \quad (14.3)$$

for $n \geqslant 1$, with the initial term $C_0 = 1$, and it is this form that they usually make their appearance. Here we shall take (14.3), and *not* (14.1), as our definition of the Catalan numbers C_n, and only later will we be able to prove that they satisfy (14.1) and (14.2). For the record, the sequence $C_0, C_1, \ldots$ is $1, 1, 2, 5, 14, 42, 132, 429, \ldots$.

14.2 Binary operations

Let $*$ be a binary operation on X, and let $a_1, \ldots, a_{n+1}$ be elements of X. The composition $a_1 * \cdots * a_{n+1}$ requires n applications of the operator $*$, and since $*$ can only be used to *combine two elements at a time* (this is what it means to say that $*$ is a *binary* operation), it is necessary to insert n pairs of brackets (or parentheses) into the composition to guarantee that it is properly defined. These brackets will tell us in which order the n operations are to be applied. For example, if $n = 2$, there are exactly two ways to do this, namely

$$\big(a_1*(a_2*a_3)\big), \quad \big((a_1*a_2)*a_3\big), \tag{14.4}$$

and if the two ways in (14.4) give the same result then we say that $*$ is *associative*. Of course, if $*$ is associative then $a_1 * \cdots * a_{n+1}$ has the same value regardless of the order in which we apply the n operations $*$. However, there are many important operations that are not associative; for example, subtraction is not associative because $1 - (2 - 3) \neq (1 - 2) - 3$. If $n = 3$ there are exactly five ways to insert the pairs of brackets, namely

$$[(a * b) * c] * d, \quad (a * b) * (c * d), \quad a * [b * (c * d)],$$
$$[a * (b * c)] * d, \quad a * [(b * c) * d].$$

Problem 14.1 Verify that these are the only possibilities when $n = 3$. How many different compositions are there when $n = 4$, and when $n = 5$?

The problem now is to find how many different ways can we insert n pairs of brackets into $a_1 * \cdots * a_{n+1}$ or, equivalently, what is the largest number of different expressions that can be obtained by inserting n pairs of brackets into the expression $a_1 * \cdots * a_{n+1}$? Let the answer to this problem (with n operations) be B_n. It is natural to define $B_0 = 1$ for the single element a_1 has a unique value. Also, it is easy to see that $B_1 = 1$, $B_2 = 2$. Now consider B_3. This is the number of ways of putting brackets into the expression $a*b*c*d$ and, as we have just seen, $B_3 = 5$. We have now shown that $B_n = C_n$ for $n = 0, 1, 2, 3$.

Theorem 14.1   *For all n, B_n is the n-th Catalan number C_n.*

Proof   Consider the situation in which the k-th occurrence of $*$ (counting from the left) is the last to be applied; then the final step in the calculation is described by the formula

$$\left(a_1 * \cdots * a_k\right) * \left(a_{k+1} * \cdots * a_{n+1}\right).$$

This situation arises in $B_{k-1}B_{n-k}$ different ways, where $1 \leqslant k \leqslant n$, so that

$$B_n = B_{n-1} + B_1 B_{n-2} + \cdots + B_{n-2}B_1 + B_{n-1}$$
$$= B_0 B_{n-1} + B_1 B_{n-2} + \cdots + B_{n-2}B_1 + B_{n-1}B_0.$$

Since $B_0 = C_0$, and both satisfy the same recurrence relation, we must have $B_n = C_n$ for all n.   □

14.3 Paths joining lattice points

A *lattice point* in the plane $\mathbb{R}^2$ is a point (x, y), where x and y are integers. For any non-negative integer n, a *path* on the interval $[0, n]$ is a sequence $(0, y_0), (1, y_1), \ldots, (n, y_n)$ of lattice points such that $y_0 = 0$ and $|y_j - y_{j+1}| = 1$, together with the segments that join consecutive pairs of lattice points in the sequence. We say that this path *joins* its initial point $(0, 0)$ to its *final point* (n, y_n). For example, the path on $[0, 8]$ in Figure 14.1 is obtained from the sequence

$$(0, 0), (1, 1), (2, 2), (3, 1), (4, 2), (5, 1), (6, 0), (7, 1), (8, 0).$$

The points (j, y_j) are called the *vertices* of the path. The segment from (j, y_j) to $(j + 1, y_{j+1})$ is called a *step*; it is a *downward step* if $y_{j+1} = y_j - 1$, and an *upward step* if $y_{j+1} = y_j + 1$. We can identify a path on $[0, n]$ by the sequence of upward and downwards steps as we move along the path from $(0, 0)$ to (n, y_n). Thus, if we agree to write

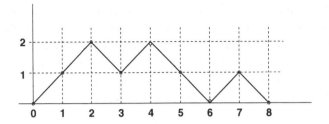

Figure 14.1 A path on [0, 8]

$+1$ when we have an upward step, and -1 when we have a downward step, each path can be identified by a sequence $(a_1, \ldots, a_n)$, where each a_j is $+1$ or -1. In fact, $a_j = y_j - y_{j-1}$ and $y_j = a_1 + \cdots + a_j$. Of course, it is clear that similar ideas and definitions apply to a path defined on any interval $[p, q]$, where p and q are integers.

Finally, we shall say that the path on $[0, n]$ given by the sequence (j, y_j), $j = 0, \ldots, n$, is a *positive path* if $y_j \geqslant 0$ for all j. These are the paths that lie on or above the x-axis.

Problem 14.2 Find a formula for the number of paths that join $(0, 0)$ to (n, k).

Problem 14.3 Find a formula for the number of positive paths on $[0, n]$.

We begin with Problem 14.2, so let $P_{n,k}$ be the number of paths that join $(0, 0)$ to (n, k). Suppose that such a path contains d downwards steps and u upward steps (in some order). Then $d + u = n$ and $u - d = k$, independently of the order, so that

$$2u = n + k, \quad 2d = n - k, \quad 0 \leqslant u \leqslant n, \quad 0 \leqslant d \leqslant n.$$

Thus *if a path from* $(0, 0)$ *to* (n, k) *exists* (it will not exist if $|k| > n$), then $-n \leqslant k \leqslant n$, and both $n - k$ and $n + k$ are even. Given these conditions such paths do exist, for we can choose any d of the n steps to be the downwards steps, and the remaining u steps be upward steps. Thus we have proved that

$$P_{n,k} = \begin{cases} \binom{n}{d} & \text{if } n = k + 2d \text{ and } |k| \leqslant n, \\ 0 & \text{otherwise.} \end{cases} \tag{14.5}$$

For example, $P_{6,2} = 15$: can the reader identify all 15 of these paths? Note that $P_{n,k} = 0$ if $n-k$ is odd; this is because after an odd number of steps to the right, we will be at an odd height from the initial position; after an even number of steps to the right we will be at an even height.

We have just seen that there exists a path from $(0, 0)$ to $(n, 0)$ if and only if n is even and, from (14.5), we see that

$$P_{2n,0} = \binom{2n}{n} = \frac{(2n)!}{n!\,n!}, \tag{14.6}$$

and, for brevity, we shall write P_{2n} instead of $P_{2n,0}$. Thus P_{2n} is the number of paths from $(0, 0)$ to $(2n, 0)$.

We can obtain a better idea of the size of P_{2n} if we can get a better idea of the size of n! We can, and this is given by *Stirling's formula* (which we shall not prove here), namely that

$$n! \sim \left(\frac{n}{e}\right)^n \sqrt{2\pi n}. \tag{14.7}$$

We are using $\sim$ here in a precise way: namely that $f(n) \sim g(n)$ as $n \to \infty$ if and only if $f(n)/g(n) \to 1$ as $n \to \infty$. It is important to realise that $f(n) \sim g(n)$ does *not* mean that $|f(n) - g(n)| \to 0$; for example, if $f(n) = n^2 + n$ and $g(n) = n^2$, then $f(n) \sim g(n)$ as $n \to \infty$, but $|f(n) - g(n)| \to +\infty$ as $n \to \infty$. If we use Stirling's theorem in (14.6) we find that

$$P_{2n} \sim \frac{4^n}{\sqrt{\pi n}}, \tag{14.8}$$

and also that $P_{2n+2} \sim 4P_{2n}$ as $n \to \infty$.

14.4 The number of positive paths on $[0, 2n]$

We recall Problem 14.3 that asks for the number of positive paths on $[0, n]$. We know that there are no such paths if n is odd, so we let Q_n be the number of positive paths on $[0, 2n]$. Our main result is that

$$Q_n = C_n, \quad n = 0, 1, \ldots, \tag{14.9}$$

so once again the Catalan numbers arise as the answer to a counting problem.

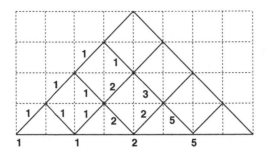

Figure 14.2 The number of positive paths

We can find Q_n for small values of n as follows. Consider Figure 14.2, where the rectangle is divided into squares by the dotted lines $x = 0$, $x = 1, \ldots, x = 8$ and the lines $y = 0$, $y = 1$, $\ldots$, $y = 4$. On each sloping edge, we write the number of ways that we can reach the left-hand vertex of that edge by travelling along a path from $(0, 0)$ to that vertex that stays on or above the x-axis. Obviously, the number attached to an edge is equal to the sum of the numbers on the edges that lead into it from the left. The numbers along the bottom of the rectangle show the number Q_n; thus $Q_0 = Q_1 = 1$, $Q_2 = 2$ and $Q_3 = 5$. As the reader can check (by completing the labelling of the edges) $Q_4 = 14$. Thus $Q_n = C_n$ for $n = 0, 1, 2, 3, 4, 5$.

We shall now find a recurrence relation for Q_n, and as this will be the same as the recurrence relation for C_n, this will prove (14.9).

Theorem 14.2   *We have $Q_0 = 1$ and, for $n = 1, 2, \ldots$,*

$$Q_n = Q_0 Q_{n-1} + Q_1 Q_{n-2} + \cdots + Q_{n-2} Q_1 + Q_{n-1} Q_0. \quad (14.10)$$

Thus $Q_n = C_n$ for all n.

Proof   We recall that after leaving its initial point $(0, 0)$, a positive path can only return to the real axis at a point of the form $(0, 2k)$. Let N_k be the number of positive paths on $[0, 2n]$ that start at $(0, 0)$, meet the real axis for *the first time* after $(0, 0)$ at the point $(0, 2k)$ and then end at $(0, 2n)$. Explicitly, this means that $y_0 = y_{2k} = 0$ but $y_j > 0$ for $1 \leqslant j < 2k$. Clearly

$$Q_n = N_2 + N_4 + \cdots + N_{2n}.$$

It is easy to find an explicit formula for N_{2k}; indeed, it is clear that $N_{2k} = A_k B_k$, where A_k is the number of paths from $(0, 0)$ to $(0, 2k)$ that stay strictly above the real axis between these two points, and B_k is the number of paths from $(0, 2k)$ to $(0, 2n)$ that stay on or above the real axis. Obviously, $B_k = Q_{n-k}$. Now let us consider a path that contributes to the count A_k. Such a path must move from $(0, 0)$ to $(1, 1)$, then stay on or above the line $y = 1$ until it reaches $(2k - 1, 1)$, and then move to $(2k, 0)$. This shows that $A_k = Q_{k-1}$. Thus $N_k = Q_{k-1} Q_{n-k}$, so that

$$Q_n = Q_0 Q_{n-1} + Q_1 Q_{n-2} + \cdots + Q_{n-2} Q_1 + Q_{n-1} Q_0$$

which is (14.10). $\square$

14.5 Photographs

Consider the following problem. We want to take a photograph of $2n$ people standing in two rows, each of n people, with one row in front of the other. We assume that no two people are of the same height, and we insist that

(1) each person in the front row must be shorter than the person standing directly behind them and
(2) the heights of the people in each row must be decreasing from left to right.

It is clear that the tallest person must stand on the extreme left position of the back row and that the shortest person must stand on the extreme right of the front row. Let us give the $2n$ people the labels $1, 2, \ldots 2n$, where the shortest person has the label 1, and so on, and the tallest person has the label $2n$. A solution to the photograph problem can now be described as a sequence of F's and B's (the front row, or the back row) as follows. We write down the integers $2n, 2n - 1, \ldots, 3, 2, 1$ in this order and then replace each integer k in this list by F if person k is in the front row, or by B if person k is in the back row. Clearly this sequence of B's and F's must start with a B and end with an F. For example, the following two expressions represent the same arrangement for a photograph of six people:

$$\begin{bmatrix} 6 & 5 & 2 \\ 4 & 3 & 1 \end{bmatrix}, \quad BBFFBF.$$

In fact, for six people ($n = 3$) there are five possible arrangements, namely

$$\begin{bmatrix} 6 & 5 & 4 \\ 3 & 2 & 1 \end{bmatrix}, \begin{bmatrix} 6 & 5 & 3 \\ 4 & 2 & 1 \end{bmatrix}, \begin{bmatrix} 6 & 5 & 2 \\ 4 & 3 & 1 \end{bmatrix}, \begin{bmatrix} 6 & 4 & 3 \\ 5 & 2 & 1 \end{bmatrix}, \begin{bmatrix} 6 & 4 & 2 \\ 5 & 3 & 1 \end{bmatrix},$$

and these can be expressed more economically as

$$BBBFFF, \ BBFBFF, \ BBFFBF, \ BFBBFF, \ BFBFBF$$

Problem 14.4 Verify that these are the only possible arrangements when $n = 3$. Find the number of possible arrangements for the photograph with 4, 8 and 10 people.

What are the necessary and sufficient conditions on a sequence of B's and F's of length $2n$ for the sequence to represent an acceptable photograph of the $2n$ people? As we have already remarked, the sequence must begin with B since the tallest person must be at the left-hand position of the back row. The next term can be either a B or an F, but if it is an F, then the third term must be a B. More generally, suppose that a person with label ℓ stands at the k-th position from the left in the front row. Then there must be at least k B's, and at most $k - 1$ F's, that have replaced the labels $2n, 2n - 1, \ldots, \ell + 1$. In short, as we progress through the sequence of B's and F's from the left, there must always be at least as many B's as there are F's. If we now rewrite the sequence of B's and F's by replacing B with $+1$ and F with -1, we obtain exactly the same set of sequences as we did for the problem of counting positive paths. The conclusion is now clear: there is a one-to-one correspondence between the positive paths on $[0, 2n]$ and the possible arrangements for a photograph; thus *there are C_n different ways to photograph $2n$ people in this way.*

14.6 An explicit formula for C_n

In this section we derive the following explicit formula for the Catalan number C_n, where $n \geqslant 1$. By definition, $C_0 = 1$.

Theorem 14.3   *For $n \geqslant 1$ we have*

$$C_n = \frac{1}{n+1}\binom{2n}{n} = \binom{2n}{n} - \binom{2n}{n+1}. \qquad (14.11)$$

Proof   We use the powerful idea of a generating function. Let

$$f(x) = \sum_{n=0}^{\infty} C_n x^n = 1 + x + 2x^2 + 5x^3 + 14x^4 + 42x^5 + \cdots .$$

The idea here is that we have collected the values $C_0, C_1, C_2, \ldots$ in *one single function*, and we can recapture these values by differentiating the function; for example, $2C_2 = f''(0)$. However, this idea will only be valid if the power series for f has a positive radius of convergence. We begin by showing that this is so. Certainly $0 \leqslant C_n \leqslant P_{2n}$, and $P_{2n+2} < 4P_{2n}$. This guarantees that the series $\sum_n P_{2n} x^n$ converges when $|x| < 1/4$ and so, by the Comparison Test, the same is true for the series for f. If we now apply standard theorems on power series, and the recurrence relation for the C_m, we find that

$$\begin{aligned}
f(x)^2 &= \left(\sum_{m=0}^{\infty} C_m x^m \right) \left(\sum_{n=0}^{\infty} C_n x^n \right) \\
&= \sum_{k=0}^{\infty} \left(\sum_{i+j=k} C_i C_j \right) x^k \\
&= \sum_{k=0}^{\infty} C_{k+1} x^k \\
&= \frac{f(x) - 1}{x} \\
&= 1 + 2x + 5x^2 + 14x^3 + \cdots .
\end{aligned}$$

This gives $xy^2 - y + 1 = 0$, where $y = f(x)$, so that

$$f(x) = \frac{1 - \sqrt{1 - 4x}}{2x}.$$

Since

$$\sqrt{1 - 4x} = 1 - 2\sum_{m=1}^{\infty} \frac{1}{m}\binom{2m-2}{m-1} x^m,$$

we see that

$$\sum_{n=0}^{\infty} C_n x^n = \sum_{m=1}^{\infty} \frac{1}{m} \binom{2m-2}{m-1} x^{m-1} = \sum_{n=0}^{\infty} \frac{1}{n+1} \binom{2n}{n} x^n.$$

Although the last few steps in this proof are technical, they should not be allowed to distract the reader from the very important idea of a generating function of a sequence. $\square$

Problem 14.5 Show that as $n \to \infty$,

$$C_n \sim \frac{4^n}{n^{3/2}\sqrt{\pi}}.$$

14.7 Why is 0! = 1

In order to work with C_n, we need to understand the factorial function. First, for any positive integer n, we have

$$\frac{d}{dt}\left(t^n e^{-t}\right) = n t^{n-1} e^{-t} - t^n e^{-t},$$

and if we integrate both sides of this equation we obtain

$$n! = \int_0^\infty t^n e^{-t}\, dt. \tag{14.12}$$

Given this, it seems reasonable to define 0! by

$$0! = \int_0^\infty t^0 e^{-t}\, dt = \int_0^\infty e^{-t}\, dt = 1.$$

In fact, we can take this idea even further and *define* x! by

$$x! = \int_0^\infty t^x e^{-t}\, dt$$

for any real x for which this improper integral converges. Now this integral converges (at 0) when $x > -1$, so we have now defined 'factorial x' (that is, x!) for all real x with $x > -1$. We have, for example,

$$(\sqrt{2})! = \int_0^\infty t^{\sqrt{2}} e^{-t}\, dt.$$

Of course, this 'new' definition agrees with the 'old' definition of n ! when n is a positive integer.

Much has been written about the integral used to define the factorial function but, more generally, the *Gamma function* is defined by

$$\Gamma(z) = \int_0^\infty t^{z-1} e^{-t} \, dt,$$

so that $n! = \Gamma(n+1)$. The notation $\Gamma(z)$, and the name *Gamma function* were introduced by Legendre in 1811, and this is a function of a *complex variable* z which is defined whenever $z = x + iy$, where $x > -1$. Better still, $\Gamma(z)$ can be extended to a function that is meromorphic throughout the complex plane $\mathbb{C}$, so we can now speak of $z!$ *for every complex number z.*

References

[1]   Ball, D. G., Squares on a square pinboard, triangles on a triangular pin-board, and hexagons on a hexagonal pin board, *Math. Gazette* LV, No. **394** (1971), 373–379.

[2]   Barbeau, E. J., *Pell's Equation*, Springer, 2003.

[3]   Beardon, A. F., Sums of squares of digits, *Math. Gazette* **82** (1998), 379–388.

[4]   Broadbent, T. A. A., Shanks, Ferguson and π, *Math. Gazette* LV, No. **392** (1971), 243–248.

[5]   Davenport, H., *The Higher Arithmetic*, Sixth Edition, Cambridge University Press, 1992.

[6]   Ehrhart E., Sur un problème de géométrie diophantienne linéare II, *J. Reine Angew. Math.* **227** (1967), 25–49.

[7]   Grimaldi, R. P., *Fibonacci and Catalan Numbers*, John Wiley & Sons, 2012.

[8]   Hardy, G. H. and Wright, E. M., *An Introduction to the Theory of Numbers*, Fifth Edition, Oxford University Press, 1979.

[9]   Keedwell, A. D., Euclid's algorithm and the money changing problem, *Math. Gazette* **92** (2008), 259–261.

[10] Mordell, L. J., *Diophantine Equations*, Academic Press, 1969.

[11] Murphy, T., The dissection of a circle by chords, *Math. Gazette* **56** (1972), 113–115 and 235–236.

[12] Reeve, J. E., On the volume of lattice polyhedra, *Proc. London Math. Soc.* **7**, No. 3 (1957), 378–395.

Index

associative operation, 105

base
  B, 59
  10, 59
binary
  arithmetic, 58
  operation, 105
bisectors, 66
Brahmagupta, 13

Catalan numbers, 104
Cayley–Hamilton theorem, 22
characteristic equation, 22
chords, 26
commutative ring, 74
continued fractions, 77
coprime integers, 2, 36
critical
  points, 79
  values, 79
cube, 90

dense, 52
diagonal, 34
digit, 54, 59
digital root, 55
dihedral
  angle, 94
  group, 83
Diophantine equations, 12
divides, 73
dynamical system, 50

edge, 31, 68
eigenvalues, 24, 46
eigenvectors, 46
Euler's formula, 31

face, 31
factor, 73
factorial function, 113
fixed points, 60

Gamma function, 114
Gaussian
  integers, 75
  prime, 75
gears, 51
generating function, 112
geoboard, 96
great circle, 33
greatest common divisor, 1

Hamming distance, 68
hyperbola, 9
hypercube, 68

integer base, 58
Intermediate Value Theorem, 92
irrational, 42
irreducible
  element, 73
  matrices, 76
  polynomial, 78

Lagrange's interpolation formula, 24, 29
lattice point, 5, 12, 35, 96, 106

linear
  equations, 22, 70
  map, 45

magic
  constant, 17
  square, 17
metric
  basis, 65
  dimension, 65
  space, 64

non-periodic motion, 51

open ball, 67
orthogonal matrix, 87

parallelograms, 39
path, 106
Pell's equation, 13
periodic, 44
  cycle, 60
  motion, 48
permutation, 83
Pick's theorem, 100
polyhedron, 102
prime

factors, 98
number, 73

quadrilateral, 81, 82
quaternion, 47

recurrence relation, 27, 104
right-handed system, 43
rotation, 52

scalar triple product, 43
semi-magic square, 17
skew
  quadrilateral, 89
  square, 89
symmetry, 81
  group, 83

Tchebyshev polynomials, 80

units, 73

vector
  product, 42
  space, 18
  triple product, 42
vertex, 31, 68, 90